Geography

for Cambridge IGCSE®

Revision Guide

Muriel Fretwell
David Kelly

Oxford and Cambridge
leading education together

OXFORD
UNIVERSITY PRESS

UNIVERSITY PRESS

Great Clarendon Street, Oxford OX2 6DP

Oxford University Press is a department of the University of Oxford.
It furthers the University's objective of excellence in research,
scholarship, and education by publishing worldwide in

Oxford New York

Auckland Cape Town Dar es Salaam Hong Kong Karachi
Kuala Lumpur Madrid Melbourne Mexico City Nairobi
New Delhi Shanghai Taipei Toronto

With offices in

Argentina Austria Brazil Chile Czech Republic France Greece
Guatemala Hungary Italy Japan Poland Portugal Singapore
South Korea Switzerland Thailand Turkey Ukraine Vietnam

Oxford is a registered trade mark of Oxford University Press
in the UK and in certain other countries

British Library Cataloguing in Publication Data

Data available

ISBN: 978-0-19-913703-9
10 9 8 7 6 5 4 3 2

Printed in Great Britain by Bell & Bain Ltd, Glasgow

® IGCSE is the registered trademark of Cambridge International Examinations.

Acknowledgements

The publisher would like to thank the following for their kind permission to reproduce photographs and
other copyright material

Cover images courtesy of Tim/Hazy Sun Images Ltd; Majority World CIC/Photographers Direct;
Marten Czamanske/Shutterstock; Vaclav Volrab/Shutterstock; Ssguy/Shutterstock; Orientaly/Shutterstock;
Maria Petrova/Shutterstock.

Illustrations by Barking Dog Art, Q2A Media, Steve Evans and Hart Mcleod.

Contents

Extra worksheets and answers to all the practice questions can be found at:
www.oxfordsecondary.co.uk/geogrg

1 World population growth

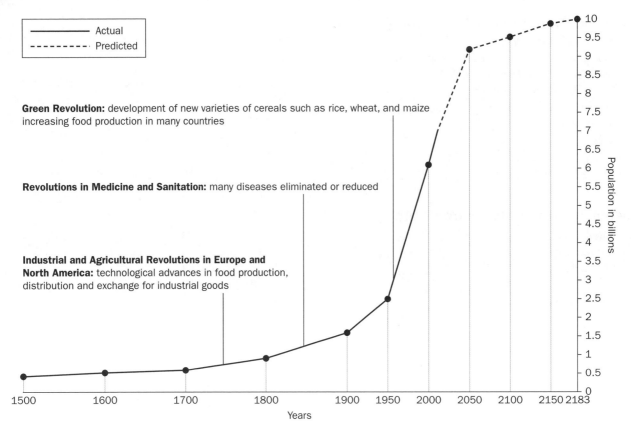

Green Revolution: development of new varieties of cereals such as rice, wheat, and maize increasing food production in many countries

Revolutions in Medicine and Sanitation: many diseases eliminated or reduced

Industrial and Agricultural Revolutions in Europe and North America: technological advances in food production, distribution and exchange for industrial goods

▲ World population growth

KEY IDEAS

→ Until 1960 the world population grew at an increasingly fast rate. The extremely rapid increase since 1800 is described as a population explosion.

→ The *rate* of growth is slowing, as the highest growth rate was 2.2 per cent in the 1960s. It had fallen to 1.1 per cent by 2011.

→ Total world population reached 7 billion in 2011, having risen from 3 billion in the 1960s. Growth is expected to reach 10 billion around 2183, before starting to fall.

→ Populations are still growing very rapidly in some countries, especially in Africa and South East Asia; these regions are expected to have more than 60 per cent of the total world population by 2050.

→ In some LEDCs, especially in sub-Saharan Africa, HIV and AIDS caused population growth rates to fall (although total populations are still rising).

→ In MEDCs in stage 5 of the demographic transition model, total populations are falling.

→ Population growth will put a great strain on resources, such as water, food, and energy supplies, as well as leading to the loss of natural vegetation and farmland for housing developments.

The natural increase or decrease of population depends on the difference between the birth and death rates.

- birth rate = the number born each year per 1000 people
- death rate = the number of people who die each year per 1000 people
- the rate of natural increase or decrease = birth rate minus death rate

(Net International migration (Chapter 3) also changes the total population of a country).

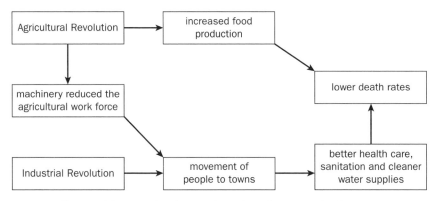

▲   The influence of the Agricultural and Industrial Revolutions on death rates

The Demographic Transition Model shows the stages in population growth that some MEDCs have experienced and other countries are expected to follow.

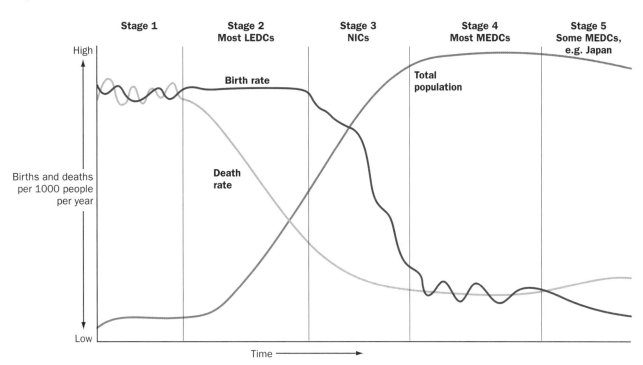

▲   The Demographic Transition Model

LEDCs have a high rate of natural increase caused by their high birth rates being considerably larger than their death rates, which have fallen rapidly (but still remain higher than those of MEDCs).
Most MEDCs have low rates of natural increase because their birth and death rates are both low.
Countries in stage 5 of the Demographic Transition Model have falling total populations because their death rates are higher than their birth rates. This is because of high life expectancies - the elderly form a high proportion of their populations (e.g. Japan and Italy).
Russia has a declining population because the birth rate is low and the death rate higher. This is because alcoholism and smoking are common and lead to more car accidents, heart attacks and general poor health. (Russia's declining population problem is made even worse because more migrants leave the country than move to live there.)

Practice questions

1. Using the information in the table, calculate the percentage rate of natural population increase for Niger (LEDC in Africa) and Ukraine (MEDC in Europe).

Country	Birth rate	Death rate	Natural population increase (%)
India (2011 est.)	21.0	7.5	1.35
Niger (2011 est.)	50.5	14.1	
Afghanistan (1950)	52.9	36.7	16.2
Ukraine (2011 est.)	9.6	15.7	
USA (2011 est.)	13.8	8.4	0.54

2. List the stages 1-5 of the Demographic Transition Model in order. For each stage, write the name of the country which has the characteristics of that stage, and the year. Choose from countries in the table above.

3. Use words from the box below to complete the table showing characteristics of each stage of the Demographic Transition Model. Some have been completed for you. You may use some more than once.

falling rapidly	falling slowly
high and fluctuating	slowly growing at first then rising rapidly
high and stationary	low and stationary
low and fluctuating	rising slowly

A summary of the Demographic Transition Model			
Stage	Birth rate	Death rate	Population total
1	high and fluctuating		
2		falling rapidly	
3			rising rapidly
4			
5			

4. Classify the following influences on high birth rates in LEDCs under the headings:

   A Cultural, social and religious **B** Demographic **C** Economic

   - children provide labour for family farms
   - children provide income to support parents
   - children give prestige to a man
   - poverty of government and citizens means little contraception provided
   - desire for a son to carry on the family name
   - high proportion of females in the population
   - illiteracy and ignorance because of lack of education
   - contraception is forbidden or discouraged
   - a large number in the child-bearing ages
   - early marriage
   - to ensure some children survive because of high infant mortality rates
   - polygamy
   - low status of women

5. The diagram shows why birth rates are falling. Suggest the reason that
each box shows and explain why it reduces birth rates.

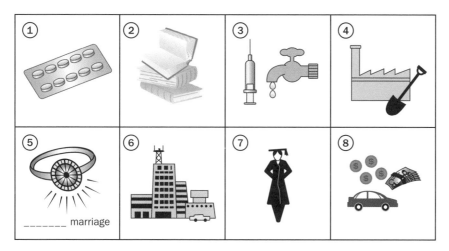

▲ Ways of reducing birth rates

6. Find six reasons (each 2 or 3 words long) to explain the fall in the
death rate which has occurred in most countries. They may be shown
horizontally or vertically.

C	F	E	D	K	L	M	A	O	P	R	T	Y	C	G	M	N	O	L	X	Z	V
L	P	B	N	M	U	O	J	L	V	D	E	M	J	Q	S	T	N	Y	L	P	A
E	T	R	N	G	R	E	A	T	E	R	K	N	O	W	L	E	D	G	E	G	C
A	H	M	Y	X	P	T	L	M	B	C	R	Y	N	O	A	N	O	V	S	T	C
N	U	I	X	P	Q	W	E	S	A	N	M	G	L	D	F	P	B	V	C	N	I
W	Q	Y	R	W	Z	X	S	A	G	F	D	C	N	M	L	K	O	P	B	C	N
A	P	S	Z	X	W	Q	U	I	R	E	D	F	G	H	J	O	L	S	N	V	A
T	O	M	S	B	E	T	T	E	R	S	A	N	I	T	A	T	I	O	N	L	T
E	A	X	C	F	G	E	W	Q	H	T	Y	U	I	O	P	L	J	K	G	D	I
R	W	F	G	E	G	H	A	X	C	V	B	N	O	P	W	D	F	S	C	O	O
S	Z	X	C	V	D	F	G	H	J	K	L	Q	R	T	Y	U	I	O	P	A	N
U	A	D	Z	X	F	G	G	H	J	K	O	L	H	R	T	Y	I	O	A	C	S
P	Q	S	D	F	G	E	U	I	P	E	V	N	M	S	Z	S	G	K	L	Q	A
P	X	B	N	A	G	K	L	M	S	F	E	T	U	I	P	Q	O	F	J	S	N
L	M	A	O	E	P	W	H	L	D	L	R	Y	V	N	M	S	J	E	D	K	D
I	M	P	R	O	V	E	D	D	I	E	T	S	A	Z	M	S	W	Q	G	L	M
E	Z	A	B	S	G	J	K	L	E	O	E	P	D	G	A	D	F	W	K	L	E
S	P	W	E	R	U	I	O	E	N	M	C	V	S	G	H	L	X	Z	C	V	D
A	X	C	V	B	M	F	H	S	K	L	W	E	T	Y	I	O	P	E	R	W	I
B	E	T	T	E	R	M	E	D	I	C	A	L	C	A	R	E	A	S	D	F	C
X	C	V	B	S	G	H	J	K	L	E	R	Y	U	I	O	P	E	D	G	S	I
C	S	F	G	H	J	L	W	R	Y	U	I	O	P	S	C	V	D	H	E	T	N
L	T	E	I	P	E	P	E	C	D	G	J	K	E	L	E	L	F	C	D	E	E
A	P	E	B	N	M	S	H	J	W	Z	A	D	F	S	C	Y	R	O	D	J	S

2 Population variations and associated problems

Over-population and under-population

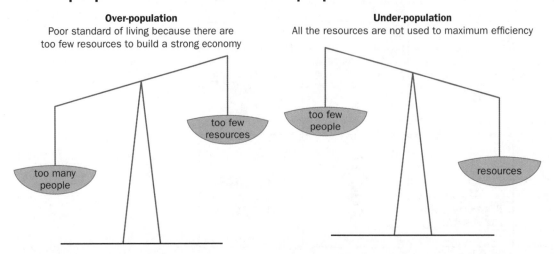

Over-population
Poor standard of living because there are too few resources to build a strong economy

Under-population
All the resources are not used to maximum efficiency

Characteristic	Under-populated country e.g. Australia	Over-populated country e.g. Bangladesh
Unemployment	Low	High
Energy and mineral resources	Many	Few
GDP (gross domestic product a measure of wealth)	Large	Small
Export earnings	High	Low
Standard of living, health and education services	High	Low
Net population movement	Immigration	Emigration
Main sector of employment	Tertiary (services)	Primary (agriculture)
Main problems caused by population imbalance with resources	Shortage of workers and foreign investors to develop and use the resources. Policy to attract immigrant workers leads to social conflict and tension in cities (e.g. Brisbane). High cost of imports of goods not produced in the country.	Shortage of food leading to over-cultivation of soils, deforestation, soil erosion and soil exhaustion, poverty, overcrowded urban areas (e.g. Dhaka), squatter settlements, water, noise and air pollution, crime.

Population structure

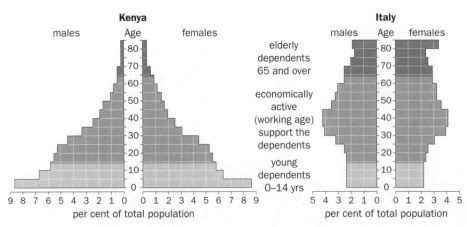

> ## KEY IDEAS
>
> → **Population structure** (the proportions or numbers of males and females in the young, middle aged and elderly age groups) is shown on a population pyramid.
>
> → The dependency ratio $= \dfrac{\text{young dependents} + \text{old dependents}}{\text{working population}} \times 100$
>
> → Countries in different stages of the Demographic Transition Model have different population structures that cause different problems that in turn need different government policies to try to solve them.

Variations in dependent populations and standards of living

Indicators of standards of living include GDP (Gross Domestic Product) per person, literacy, energy consumption, and calorie intake: these are highest in MEDCs. Financing the needs of an increasing dependent population results in lower living standards for the economically active in any country and reduces the amount available for developing the country.

	Problems	Main needs	Solutions
LEDCs	Many young dependents	More health care, schools and food supplies. More jobs for the increasing workforce.	Policies to reduce birth rates but governments have inadequate financial resources to provide enough family planning and other needs
MEDCs e.g. Japan	Many elderly dependents	More residential homes, social support and medical care. More workers to maintain industries and support the dependents	To finance the needs of the elderly, Japan increased taxes, raised the state pension age and started an insurance scheme for workers to cover the costs needed later in life. The elderly share the costs of their care. To increase the workforce Japan encouraged immigration and allowed pensioners to continue working. The alternative is to reduce services.

Mexico's changed population policies in response to changing population problems		
Date and reason	**Policy**	**Result of policy**
1936 to increase the workforce. (Population growth rate about 0%). Stage 1 of the Demographic Transition Model)	A law was passed to encourage marriage, child bearing, women's health and immigration	Very rapid growth, mainly achieved by immigration
1947 to further increase the workforce, as workers were emigrating, especially to the USA	A new law promoting greater immigration	Rapid growth resulted from a very high fertility rate and falling death rate. (now Stage 2 of the Demographic Transition Model)
1974 population rising too quickly to allow economic development	Women given equal rights to family planning and work. Aims spread by education and the media.	Growth rate slowed from 3.5% to 1.5% by 2000.

Population density and distribution

KEY IDEAS

→ **Population density** = $\dfrac{\text{total population}}{\text{area}}$

→ Population distribution is *how* the population is spread in an area.

		Reasons for low population densities	
	Physical		**Human and Economic**
Hot deserts	Too little water		Lack of transport routes
Marshy areas	Too much water		Difficult transport, few suitable settlement sites
High or steep mountain slopes	Too steep and too cold for arable farming at high altitudes		Difficult transport, few suitable settlement sites
High latitudes	Too cold for agriculture. Frozen ground and long periods of darkness make other activities difficult		Snow, ice and swampy ground in the summer make transport difficult
Other areas	Infertile soils (where no other economic activity is possible)		Lack of minerals or other resources for development

Practice questions

Which **one** statement in each of the following groups is **correct**?

1. Foreign investors are not attracted to under-populated countries because:
   a) the population is poor
   b) living standards are low
   c) the domestic market is small
   d) workers have low levels of literacy

   Under-populated countries import foodstuffs and manufactured goods because:
   a) they do not have the raw materials to manufacture them
   b) they have a shortage of workers to produce them
   c) they are cheap to import
   d) they cannot afford fertiliser and machinery

   Under-populated countries export minerals and agricultural products because:
   a) they do not have the skilled workers to turn them into manufactured goods
   b) it is cheap to import goods made from them
   c) they are surplus to the population's needs
   d) they have a shortage of suitable land for industry

2. Calculate the population densities of Australia and Bangladesh.

Country	Australia	Bangladesh
Population total	21 700 000 (2011)	158 000 000 (2011)
Area	7 617 930 km²	147 570 km²
Population density	_____ per km²	_____ per km²

3. Use the diagram to answer the following questions.

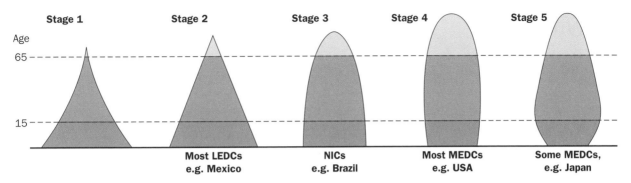

Stage 1 | Stage 2 | Stage 3 | Stage 4 | Stage 5

Age
65
15

Most LEDCs
e.g. Mexico

NICs
e.g. Brazil

Most MEDCs
e.g. USA

Some MEDCs,
e.g. Japan

a) Write the stages of the demographic transition model as a list. For each, write any the following descriptions that apply to the shape of the population pyramid it represents:

slightly convex sides, triangular, concave sides, very convex sides, straight sides (thins slowly from the base upwards), rapidly narrows from 15 years to the base

b) Identify which stage has:

i) the highest proportion of elderly dependents

ii) the highest proportion of young dependents

iii) the lowest proportion of elderly dependents

iv) the lowest proportion of young dependents

v) the lowest number of children to grow up to become workers, so a worse dependency ratio in the future,

vi) a rapidly growing population

vii) a rapidly shrinking and ageing population

viii) the highest birth rate (two possible answers)

ix) the lowest birth rate

x) the lowest life expectancy

xi) the lowest death rate.

4. Calculate the dependency ratio for Japan.

Population under 15 (thousands)	Population 65 and over (thousands)	Population 15 – 64 (thousands
16 500	31 900	80 910

5. Explain the different ways in which urbanisation leads to reduced birth rates.

6. Briefly compare the population structures of Kenya and Italy (page 6) in a table.

7. List as many reasons as you can think of for areas having high population densities. Arrange then under two headings - Physical reasons and Human and Economic reasons

3 The effects of migration and HIV/AIDS

The influence of migration on population growth rates

Migration is the movement of people from one place to another.

| source area for emigrants population decrease | — migration → | destination area for migrants population increase |

KEY IDEAS

→ Net migration for an area is calculated by:
number of immigrants − number of emigrants

→ If more people come in than leave, the net migration is a positive figure. If more leave than come in, it is a negative figure.

→ The population growth (or decline) of an area is natural change + net migration i.e.:
(birth rate − death rate) + (number of immigrants − the number of emigrants)

Types of Migration

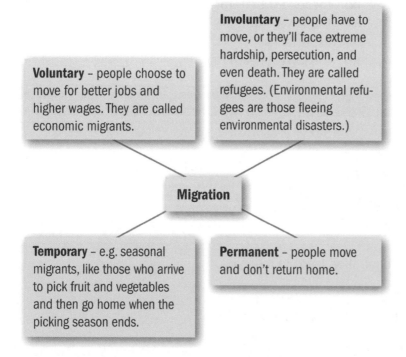

Voluntary - people choose to move for better jobs and higher wages. They are called economic migrants.

Involuntary - people have to move, or they'll face extreme hardship, persecution, and even death. They are called refugees. (Environmental refugees are those fleeing environmental disasters.)

Migration

Temporary - e.g. seasonal migrants, like those who arrive to pick fruit and vegetables and then go home when the picking season ends.

Permanent - people move and don't return home.

Migrations are either internal (within the country), or international. In LEDCs and MEDCs the largest internal migrations are rural to urban, but in MEDCs urban to rural migration is also important. Urban to urban and rural to rural movements also occur.

Most migrants are young adults, especially males. Migrations can be permanent, seasonal or daily. The movements are caused by push and pull factors.

Push factors include:

- unemployment
- low wages
- poor educational opportunities
- poor health care
- poor standards of living
- war, crime and persecution
- drought and famine
- natural disasters, such as floods and volcanic eruptions.

Pull factors include:

- seeking employment
- higher wages than at home
- better health care
- better educational opportunities
- better standards of living
- to be near friends or family
- lower levels of crime
- safety from conflict and persecution
- to be near entertainment and leisure facilities.

Most international migrations are to neighbouring countries and from LEDCs to MEDCs. An example of this is the high level of migration from Mexico (the country with the most emigrants in 2010) to the USA (the country with the most immigrants – 43 million of its citizens were born in another country).

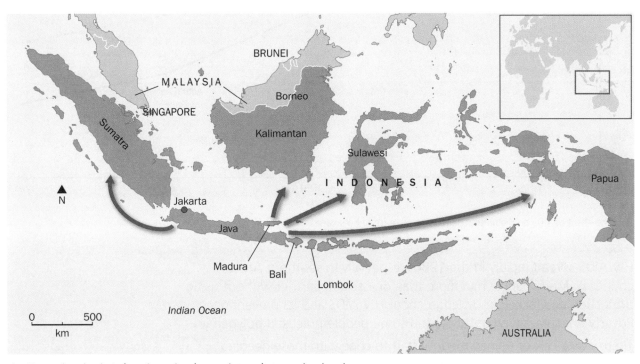

▲ Transmigration in Indonesia: an involuntary internal economic migration

The transmigration programme attempted to reduce the problems of Indonesia's over and under-populated areas. Two and a half million people have been moved from the over-populated islands to under-populated islands.

The impact of immigration on the population of the UK

There has been large-scale immigration to the UK since the mid-1990s, especially of refugees and economic migrants from LEDCs and from Eastern Europe where fertility rates are traditionally higher than in the UK. The number of births to women born outside the UK has risen every year since 1995. More than a quarter of the babies born in the UK in 2010 had mothers who were born in other countries. (Poland was the main country of origin for mothers and Pakistan of fathers.) The total UK population rose from 57.8 million in 1991 to over 62 million in 2010. This rapid growth has put health (especially maternity) services, housing provision and other services under great strain in areas where a lot of immigrants have settled.

The impact of HIV/AIDS

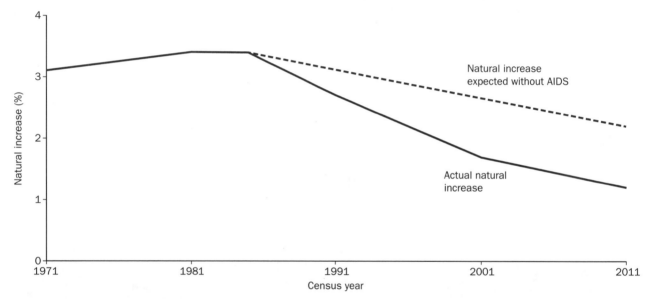

▲ The effect of AIDS on the natural population increase of Botswana.

HIV/AIDS spread rapidly in the 1980s, especially in southern African countries. Many people had more than one sexual partner without using protection and lacked knowledge about HIV/AIDS and its prevention. Poverty encouraged prostitution and some people practised polygamy.

There was a lack of trained medical staff to cope with those needing treatment. By 2002 Botswana's death rate was seven times higher than it would have been without AIDS, life expectancy had fallen from 72 to 34 and the population growth rate was only 1.1%, a third of what it was before AIDS and much lower than in most African countries.

AIDS is slowly being brought under control; Botswana's population growth rate rose to 1.65% in 2011 as a result of the antiretroviral drug programme and widespread advertising of the need to use condoms. AIDS has left countries with a shortage of middle aged parents and workers, many orphans and not enough care workers or funds to provide help for all in need.

Practice questions

1. Make a list of the difficulties international migrants might face
    a) before leaving their home country,
    b) in their host (destination) country.

2. List **(a)** the benefits and **(b)** the possible problems of immigration for the host country.

3. How might the source country or area benefit from emigration?

4. List the problems caused in rural areas by rural to urban migration.

5. Classify the examples of migration by the type of migration they represent. Use any of the words in the box below that apply to each example:

temporary   permanent   voluntary   involuntary (forced) rural
urban   internal   international   economic   environmental   commuting

Examples
    a) From the countryside to live and work in a city.
    b) From the countryside to work in a nearby city and return home at night.
    c) From Botswana to work for a number of years in the gold mines of South Africa.
    d) Retiring from work in a city to live in the countryside.
    e) Moving during the 'Arab Spring' uprising in Libya in 2011 to neighbouring countries.
    f) Moving away from the area damaged by the Japanese tsunami in 2011.
    g) Maasai people moving within Kenya with their animal herds to find pasture in wetter areas during the dry season.

6. Suggest two reasons why urban population pyramids show a lower percentage of elderly people than rural population pyramids.

4 Settlement hierarchy and pattern

Hierarchy of settlements

> **KEY IDEAS**
>
> → A **hierarchy of settlements** is a list of settlements in order of population size and the number and range of services provided.
>
> → **High order settlements** are larger, fewer in number, spaced further apart, and with a wider range of services.
>
> → **Low order settlements** are smaller, more in number, more closely spaced and with a small range of services.
>
> → **Urban** means associated with towns and cities.
>
> → **Rural** means associated with the countryside and villages.
>
> → **Services** (functions) are anything that is provided in a settlement for the population, including goods that can be bought in shops and other retail outlets.

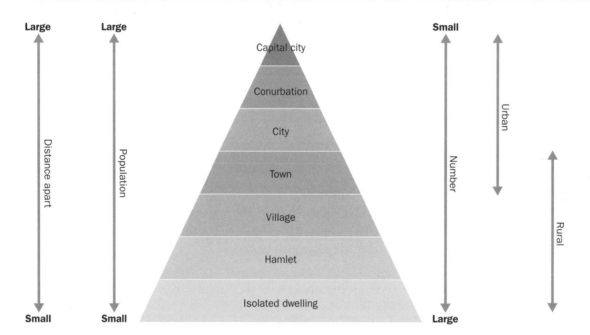

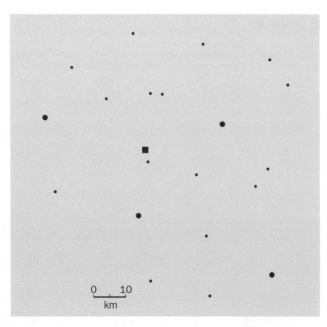

Hierarchy of services

KEY IDEAS

→ A **hierarchy of services** puts the services of a settlement or area in rank order of importance, usually based on the population needed to support the service and the frequency of use.

→ **Low order services** have a small threshold population, such as a local shop or a primary school, usually in large numbers. They are found in most settlements.

→ **High order services** have a large threshold population, such as a department store or a university, usually in small numbers. They are usually only found in the larger settlements.

→ **Convenience shops** sell goods bought almost every day from local shops, such as bread and milk.

→ **Comparison shops** sell goods not bought every day. The shopper visits more than one store to look at different prices and quality for items such as furniture or shoes.

Sphere of influence

KEY IDEAS

→ A **sphere of influence** is the area served by a settlement or service.

→ **Range** is the maximum distance that people are prepared to travel to access a particular service.

→ **Threshold population** is the minimum number of people needed to provide a large enough demand for a service.

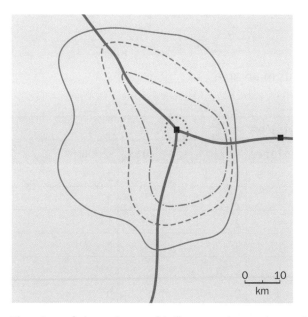

Key
— Main road
— Hospital
--- Furniture store
—·— Secondary school
······ Local shop
■ Town

0   10
km

The size of the sphere of influence depends on the:
- size of the settlement and the services provided
- population density of the area
- wealth of the people in the area
- transport facilities
- competition from other settlements.

Settlement pattern

KEY IDEAS

→ **Settlement pattern** is the shape that a settlement forms on the map and how clustered or scattered the settlement is.

→ **Nucleated** settlements have houses clustered together as villages, with fewer isolated dwellings. The shape of the villages is compact and more square or circular. They develop at cross roads, bridges, good defensive points, and where there are mineral resources. People can enjoy the social benefits of living close to their neighbours. They have easy access to services like shops and schools.

→ **Dispersed** settlements are scattered, isolated dwellings and small hamlets with few villages. They develop where the agricultural land is poor and where people need large areas of land for things such as grazing. It would be impossible to live in a village and still be within travelling distance of agricultural land.

→ **Linear** settlements are in long thin rows, often along roads or tracks. They develop along a road or track for transport, near to an area of farming land, at right angles to the road. The settlements may be along a river or a line of springs for water supply or along a valley floor avoiding steep valley sides.

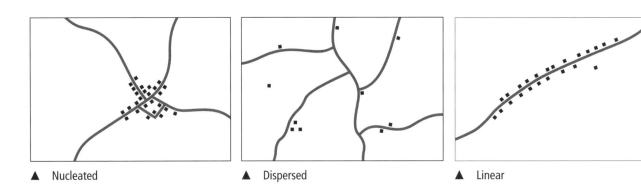

▲ Nucleated   ▲ Dispersed   ▲ Linear

Practice questions

1. The following table gives information about settlements in an area. Study the table and answer the following questions.

Settlement	Population	Number of services		
		Convenience stores	Comparison stores	Doctors
A	12 500	12	6	2
B	11 900	16	9	4
C	5300	5	2	1
D	5100	4	3	0
E	5000	5	2	0
F	800	1	0	0
G	800	1	0	0
H	700	2	1	0
I	700	1	0	0
J	750	1	0	0

a) How many levels are there in the hierarchy of settlements?
b) Which settlement is highest in the hierarchy? Explain your choice.
c) Which settlement is lowest in the hierarchy? Explain your choice.
d) How big does a settlement need to be before it has a doctor?
e) Suggest the size of population needed to support a convenience store.
f) Of the three services shown in the table, which is the highest order service? Explain your answer.

2. The following map shows the location of the settlements in the table.

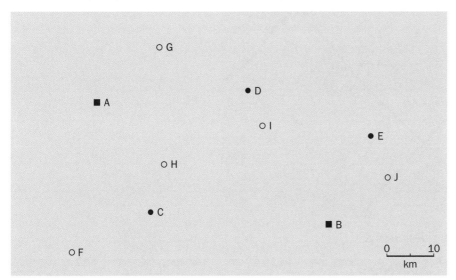

a) Make three copies of the map. On the first copy mark the spheres of influence of settlements A and B. On the second copy mark the spheres of influence of settlements C, D, and E. On the third copy mark the spheres of influence of settlements F, G, H, I, and J.
b) What is the greatest distance that people need to travel to get to a convenience store?
c) What is the greatest distance that people need to travel to get to a doctor?
d) What is the term used to describe the greatest distance that people travel to access a service?

3. For each of settlements A, B, and C shown below, describe the settlement pattern and give reasons for its location.

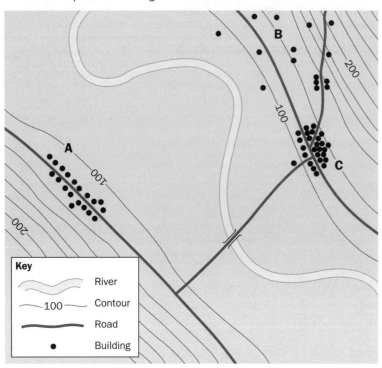

4. Match the terms with their definitions.

	Term		Definition
A	Comparison goods (comparison store)	1	Goods not bought every day. The shopper visits more than one store to look at different prices and quality of items such as furniture, shoes, or clothes.
B	Function	2	Goods bought almost every day from local shops, such as bread and milk.
C	Urban	3	(a) Another word for service, or (b) the purpose of a settlement. For example, mining or industry.
D	Hierarchy of services	4	Placing the services of a settlement or area in rank order of importance, based on the population needed to support the service and the frequency of use.
E	Hierarchy of settlements	5	A list of settlements in order of population size, number and range of functions and importance.
F	Sphere of influence	6	Services with a large threshold population, such as a department store or a university, usually in small numbers. Usually only found in the larger settlements.
G	High order settlements	7	Settlements higher up the hierarchy, such as large cities, which are fewer in number, spaced further apart, and with a wider range of services.
H	Convenience goods (convenience store)	8	Services with a small threshold population, such as a local shop or a primary school, usually in large numbers. Found in most settlements.
I	Low order settlements	9	Settlements lower down the hierarchy, such as small villages, which are higher in number, more closely spaced and with a small range of services.
J	Range	10	The maximum distance that people are prepared to travel to access a particular service.
K	Rural	11	Associated with the countryside and villages.
L	Service	12	Anything that is provided in a settlement for the population. This includes goods that can be bought in shops and other retail outlets such as food, petrol or clothing, businesses like hairdressers which are sometimes called retail services, as well as public services like schools, hospitals, government, police, water, and electricity.
M	Low order services	13	The area served by a settlement or service.
N	Threshold population	14	The minimum number of people needed to provide a large enough demand for a service.
O	High order services	15	Associated with towns and cities.

Site, situation and growth of settlements

Site and situation

> ### KEY IDEAS
>
> → **Site** is the land the settlement is built on.
> → **Situation** means the position of a settlement in relation to the surrounding area.

Factors influencing the sites and development of rural settlements

You should be able to describe the effect of the following factors:

- Agricultural land use
- Relief: altitude, gradient, aspect
- Soils
- Water supply
- Drainage and flooding
- River crossings
- Natural resources such as minerals
- Development of tourism
- Accessibility

These factors influence the site, situation and pattern of settlement on the following map and cross section.

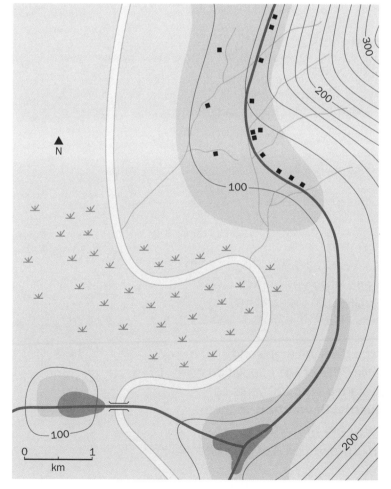

Key

- River
- Contours (metres)
- Road
- Village
- Dwelling
- Marsh
- Cultivation

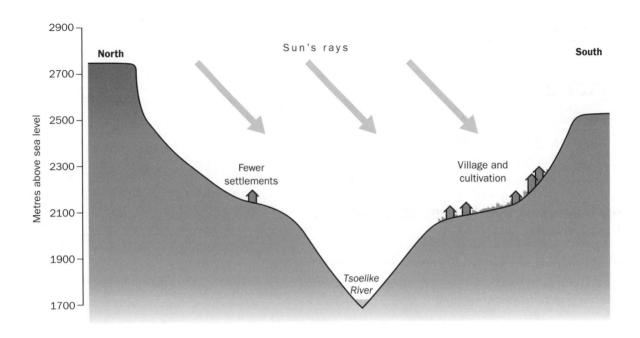

Factors influencing the size, growth and functions of urban settlements

You should be able to describe the following factors which have allowed rural settlements to grow into larger towns and cities:

- Nodal points (route centres)
- Agricultural centres
- Ports
- Administrative towns and cities

Problems and changes in rural settlements

You should be able to describe the following problems:

- Low wages and lack of jobs
- Depopulation
- Ageing population
- Decline of services
- Rising property prices

Urbanisation

KEY IDEAS

→ **Urbanisation** is the growth of towns and cities leading to an increasing percentage of the population living in urban areas.

Urbanisation took place earlier in the richer countries in Europe and North America and in MEDCs. Today more than 90% of the population in MEDCs live in towns and cities. Urbanisation started over 200 years ago when these countries went through the **Industrial Revolution**. In these countries urbanisation is now either very slow or has stopped.

The LEDCs in Africa and South East Asia have much lower levels of urbanisation because industrialisation took place later.

You should know the reasons for urban growth, including:

- overall population growth
- rural–urban migration
- increasing numbers of people working in secondary and tertiary industries, which are concentrated in urban areas.

KEY IDEAS

→ **Conurbations** have formed when cities have grown outwards and have merged with other towns and cities.

→ **Counter-urbanisation** is the movement of population from towns back to rural areas, mainly in MEDCs. You should know the reasons for counter-urbanisation.

→ **Mega-cities** are cities with populations of over ten million, including extremely large conurbations such as Tokyo-Yokohama.

Practice questions

1. For the map below,
   a) describe the settlement pattern.
   b) describe the site and situation of the settlements.

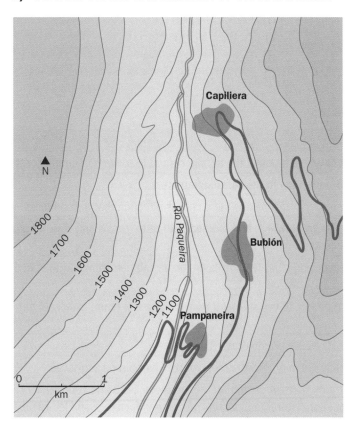

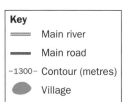

Key
≋ Main river
━ Main road
−1300− Contour (metres)
⬤ Village

2. The map below shows four settlements, A, B, C, and D.

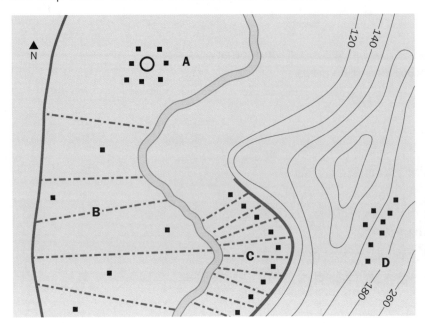

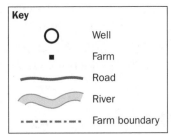

Key

◯	Well
▪	Farm
⎯⎯⎯	Road
≈≈≈	River
·–·–·–	Farm boundary

For each settlement,

a) describe the settlement pattern.

b) explain why the settlement has grown.

3. For a rural area or a village which you have studied, explain how each of the following factors has affected its growth. Give local information about your chosen rural area or village.

   a) Relief (altitude, gradient, aspect)
   b) Water supply
   c) Transport routes
   d) Agriculture

4. For a rural area which you have studied, explain how it has been affected by the following problems:

   a) Low wages and lack of jobs
   b) Depopulation
   c) Ageing population
   d) Decline of services
   e) Rising property prices

5. Match the terms with their definitions.

	Term		Definition
A	Industrial Revolution	1	A small settlement where many of the people commute to work in another settlement.
B	Millionaire city	2	The position of a settlement in the surrounding area.
C	Aspect	3	When used in geography, this means the compass direction that a slope faces.
D	Dry point	4	Cities that have grown outwards and merged with other towns and cities to produce very large settlements.
E	Counter-urbanisation	5	The growth of towns and cities leading to an increasing percentage of the population living in urban areas.
F	Dormitory village	6	How easy it is for people to travel to and from a particular point, usually in terms of time travelled.
G	Entrepôt	7	Higher points in otherwise poorly drained areas.
H	Situation	8	A port where goods are imported then re-exported without paying taxes.
I	Wet point	9	The process that took place in certain MEDCs between 1750 and 1900 which changed the economy from agricultural to industrial, developed factories, and led to urbanisation.
J	Mega-city	10	A city with a population of over 10 million people.
K	Hinterland	11	An area linked to the port from which goods are exported and to which goods are imported.
L	Site	12	The land that a settlement is built on.
M	Conurbations	13	The movement of population from towns back to rural areas.
N	Urbanisation	14	A city with a population of over 1 million people.
O	Accessibility	15	A site with reliable sources of water from rivers, springs, and wells in an otherwise dry area.

Urban land use

Concentric zone model

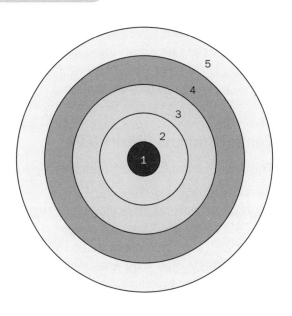

The CBD develops at the original growth point which was at the intersection of major roads. Beyond the CBD is a manufacturing zone. New immigrants moving into a city move into inner city areas with cheap housing, close to sources of employment. Housing quality and social class increase with increasing distance from the centre. Increasing affluence and developing public transport allow people to live long distances from their places of work. The area next to the expanding CBD is the transition zone where residential areas are changing to commercial use.

Sector model

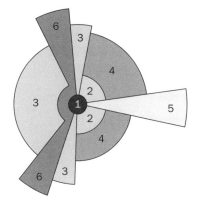

Key
1 CBD
2 Light manufacturing
3 Low-class residential
4 Middle-class residential
5 High-class residential
6 Heavy manufacturing

This is based on transport routes and the idea that certain types of land use repel each other. Industry develops along major roads or rivers. Manufacturing industry and high-class residential areas are never next to each other.

Models of cities in LEDCs

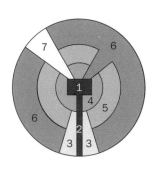

Key
1 CBD
2 Commercial
3 High-class residential
4 Better residences
5 Active improvement of houses
6 Recent squatter housing
7 Industry

In many cities in LEDCs, the high-class residential areas are close to the centre and the low-class residences are at the edge of the city – the exact opposite of cities in MEDCs.

The CBD is based on the old, colonial centre and has a sector of shops and offices leading from it along a major transport route. Either side of this are high-class residential sectors. These contain open areas, parks, homes for the upper and middle classes and amenities such as good schools.

The other residential areas are based on concentric zones. Recent squatter settlements are found on the outskirts while closer to the centre, housing conditions are better and there is older, more established squatter housing. Manufacturing tends to be scattered throughout the city, although there may be industrial sectors along transport routes.

Central Business District (CBD)

This includes the following features:

- Government buildings.
- High order retail services such as department stores in the middle of the CBD and highly specialist shops on the outskirts of the CBD.
- Offices, including major company headquarters.
- Theatres, hotels, and restaurants.
- Old, historic buildings.
- Multi-storey buildings, developed in response to the high land values.
- Public transport services including buses and underground railways.
- Few residents – the number of people in the CBD at night is low.
- Zoning of different functions in different parts of the CBD. This is because certain shops, such as shoe shops, are better next to shops of the same type for comparison shopping. Businesses like banks and legal services also prefer to be next to each other for business contacts.
- Vertical zoning – retail on the lower floors, offices on the upper floors.
- High numbers of pedestrians.
- Pedestrianised areas.

Reasons for the development of the CBD

- In many towns and cities, the CBD represents the original growth point of the settlement.
- The CBD was also the point where roads from the outskirts converged. This made the area the most accessible area of the town. This in turn made it a very desirable place for services like retailing.
- Land prices became high and only certain services could afford to locate there. Buildings grew tall to make the best use of the expensive land.

Other types of urban land use

- High-density housing
- Low-density housing
- Apartments (Flats)
- Shanty houses
- Open spaces
- Transport routes
- Industrial areas

Practice questions

1. Study the following model of urban land use.

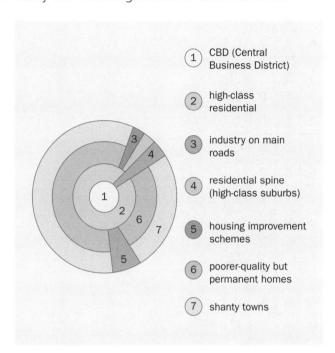

Does the diagram represent a city in an MEDC or a city in an LEDC?
Give reasons for your choice.

2. The diagram opposite shows the quality of life in cities in LEDCs and MEDCs.

   a) Describe the differences in quality of life between cities in LEDCs and cities in MEDCs.

   b) Give reasons for the differences which you have described.

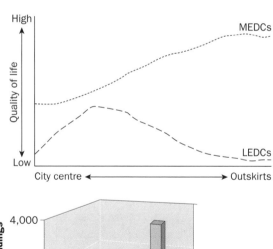

3. The diagram opposite shows how the height of buildings has changed in Shanghai, (China) between 1980 and 2005.

   a) Describe the heights of the buildings in Shanghai in 1980.

   b) How has this changed by 2005?

   c) Suggest in which zone of the city the buildings of over 30 storeys will be found and explain your choice.

4. Choose a town or city which you have studied. Describe the land use of the city and explain how it fits (or does not fit) models of urban morphology.

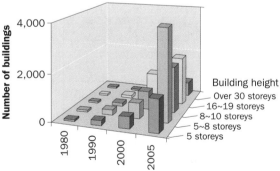

5. Find the words which match the following descriptions. They may be shown horizontally or vertically.

A	C	B	E	F	B	R	O	W	N	F	I	E	L	D
L	Z	P	S	B	M	D	G	K	L	O	P	R	Y	C
L	A	C	S	H	A	N	T	Y	P	R	T	R	K	W
Z	G	E	W	Y	T	I	U	O	P	A	S	E	D	M
O	R	X	S	V	B	N	E	G	M	G	R	S	D	S
P	E	P	A	Z	S	F	H	H	K	L	T	I	O	Q
C	T	D	B	N	F	M	S	E	A	L	E	D	S	F
P	A	E	U	Y	O	P	E	T	M	D	L	E	K	S
O	I	R	E	D	F	S	J	T	L	C	X	N	Z	A
J	L	H	F	S	O	P	E	O	W	A	V	T	D	S
A	Q	W	T	Y	U	I	E	R	U	I	O	I	S	S
D	T	R	A	N	S	I	T	I	O	N	W	A	X	O
S	P	W	M	D	F	X	V	D	B	N	U	L	I	A
A	S	X	V	N	D	M	J	L	R	O	E	S	G	J
I	W	G	R	E	E	N	F	I	E	L	D	X	U	D

a) This slum dwelling is found in LEDCs.
b) This site was built on, but the buildings have been knocked down for re-development.
c) Another name for a housing area.
d) In this part of the city a racial group lives in relative poverty.
e) This site was not built on before.
f) On the edge of the CBD, this zone has poor buildings and high rates of crime and social problems.
g) This word just means shopping.

6. Match the terms with their definitions.

	Term		Definition
A	Low-density housing	1	The central area in a town or city, with the highest land prices, greatest accessibility, and a concentration of big shops and offices.
B	Central Business District (CBD)	2	A housing area.
C	Greenfield site	3	The distribution of land use types in an urban area – the same as urban morphology.
D	High-density housing	4	A site not previously built on.
E	Brownfield site	5	Housing with fewer dwelling units per square kilometre.
F	Retail	6	Shopping
G	Residential area	7	The concept that the centre and outskirts of the CBD will have different and distinctive land use types.
H	Core and frame	8	A site that has been previously built on and could be re-developed.
I	Urban structure	9	Housing with a large number of dwelling units per square kilometre.

Problems of urban areas

Some of the problems listed below apply particularly to the CBD while other problems apply to the whole city. Environmental problems are dealt with in the next chapter.

Problem	Solution
Decline of the CBD in cities in MEDCs Issues include: • congestion and poor accessibility • lack of parking space • high land prices • retailers leaving (smaller shops which can no longer afford the high rents; shops such as DIY, furniture and carpet shops have often moved to brownfield sites; major department stores and hypermarkets have moved to greenfield sites on the outskirts), leading to empty derelict buildings • decentralisation of companies and administration to the outskirts.	**Pedestrianisation** • Allow a more safe, relaxed environment. • Less air and noise pollution from vehicles. **Shopping malls** • Undercover shopping areas. • Air-conditioned malls. • Cafés and small restaurants. **Visual improvements** • Flower beds, seated areas, trees and hanging baskets. • Pavement cafés and bars are introduced. **Transport improvements** • Underground railways
Security in the CBD, especially in evenings, in cities in MEDCs and LEDCs • High crime rates • Litter • Graffiti	• Patrols by police or by private security firms. • Closed circuit TV is a deterrent to pick pockets and shoplifters.
The twilight zone in cities in MEDCs This is the transition zone on the edge of the CBD. It can also be an area of decline and can suffer from: • derelict land and buildings • high rates of crime and social problems.	• Re-development by city governments.
Crime and racial conflict in cities in MEDCs and LEDCs • High levels of poverty. • Development of ghettos in inner cities in MEDCs. • Informal settlements in LEDCs.	• Providing social facilities such as sports clubs. • Job creation schemes to provide employment. • Special projects that bring communities together. • Zero tolerance on crime. • Ensuring adequate policing on the streets. • Providing language lessons for immigrants.
Squatter settlements in cities in LEDCs • Residents do not own the land or have a legal right to occupy the land and could be evicted at any time. • Houses are not weatherproof and can be cold in winter. • No proper sanitation and water supply, leading to diseases such as cholera. • No refuse collection. • No electricity supply, or if there is a supply, it is illegal. • Location on the outskirts means that there is no local employment. • Long journeys to the central areas of the city for work and little public transport. • Extreme poverty and high unemployment. • Extreme overcrowding with families living in one or two rooms. • High levels of crime and drug and alcohol abuse.	• Low-cost housing schemes. • Provision of piped water. • Provision of electricity. • Provision of sewers. • Provision of refuse collection. • Self-help schemes which provide groups of people with the materials to build proper houses.

Housing shortages in MEDCs These are caused by: • older properties nearer to city centres requiring renovation or renewal • a population increase through immigration and natural growth • property prices being too high for those who are unemployed or on low wages.	• Slum clearance schemes. • Older housing replaced by blocks of flats or new houses in the suburbs. • New towns built in the countryside. • In Japan, land reclaimed from the sea to build houses on.
Traffic congestion in cities in MEDCs and LEDCs • Ancient cities built long before the need for mass public transport. • Increased use of private cars. • Large numbers of commuters so that many of the trains and buses that carry these people are not needed during the rest of the day. • People visiting the CBD for sightseeing, shopping or entertainment. • People passing through the city on their way to other places.	• Underground railways • Bus lanes • Congestion charge • Electronic ticketing • Integrated transport policies • Ring roads • Traffic lights (robots) • Tidal flow • Trams • Roundabouts (circles) • Park and ride schemes
Urban sprawl and pollution – see chapter 8.	

Practice questions

1. The following diagram shows poor quality housing in cities in MEDCs and LEDCs.

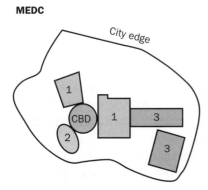

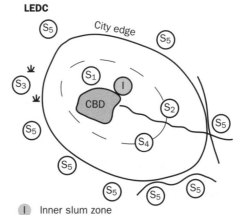

1 Inner-city areas, a mixture of redeveloped nineteenth-century slums and modern high-rise flats.

2 Inner-city transitional areas; former villas have become bedsits and multi-let homes

3 Outer-city council estates, many very deprived

I   Inner slum zone
S   Squatter settlement:
    (S₁) on temporary site
    (S₂) on a river bed
    (S₃) on a marsh
    (S₄) on a hillside, subject to landslides
    (S₅) new shanty towns on edge of city

a) Describe the locations of poor quality housing in MEDCs.
   What problems will these areas face?
b) Describe the locations of poor quality housing in LEDCs.
   What problems will these areas face?

2. The following diagram shows the core and frame of the CBD.

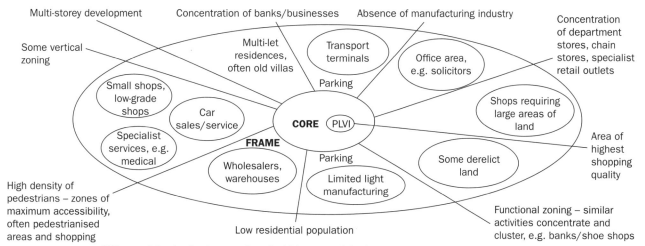

Multi-storey development Concentration of banks/businesses Absence of manufacturing industry Concentration of department stores, chain stores, specialist retail outlets

Some vertical zoning

Multi-let residences, often old villas

Transport terminals

Office area, e.g. solicitors

Parking

Small shops, low-grade shops

Car sales/service

CORE (PLVI)

Shops requiring large areas of land

Area of highest shopping quality

Specialist services, e.g. medical

FRAME

Parking

Some derelict land

High density of pedestrians – zones of maximum accessibility, often pedestrianised areas and shopping

Wholesalers, warehouses

Limited light manufacturing

Low residential population

Functional zoning – similar activities concentrate and cluster, e.g. banks/shoe shops

PLVI = peak land value intersection: the highest rated, busiest, most accessible part of CBD

Make a list of the services in the core of the CBD and another list of the services in the frame of the CBD.

3. A company has its head office in the CBD of a city. It needs new bigger offices and decides to re-locate to the edge of the city. List the reasons for this decision. Include disadvantages of the CBD and advantages of a location on the edge of the city.

4. The following maps show the highest concentrations of the homes of different ethnic groups in London, UK.

CHINESE   INDIAN   PAKISTANI   BANGLADESHI

a) Suggest reasons for the concentrations of the different ethnic groups.
b) What name is given to an area of a city where a concentration of an ethnic group live in relative poverty?

5. For an urban area which you have studied, describe the problems it has experienced and the improvements which have been made.

6. Match the terms with their definitions.

	Term		Definition
A	Twilight zone	1	Travelling to and from work, usually from home, on a daily basis. This term usually implies that the journey is not extremely short.
B	Slum	2	Part of an urban area where there is a concentration of a single racial group living in relative poverty.
C	Park and ride	3	System used to reduce traffic congestion where car parks are provided on the edge of the built-up area and there are buses to the centre.
D	Shanty	4	A slum dwelling which is a temporary or poorly-built construction, often on land that does not belong to the householder.
E	Commuting	5	Housing which is considered to be below acceptable standards.
F	Transition zone	6	The edge of the CBD which is in the process of change, either because the CBD is expanding or because the area is going into decline.
G	Ghetto	7	Problem area in the zone on the edge of the CBD. It has been badly affected by change and may have derelict land and buildings and high rates of crime and social problems.

7. Complete the following crossword.

Across

3. This zone is on the edge of the CBD which is in the process of change, either by the CBD expanding or by the area going into decline. (10)
5. This is housing which is considered to be below acceptable standards. (4)
7. Car parks are provided on the edge of the built-up area and there are buses to the centre. (4/3/4)
8. In this part of the city there is a concentration of a single racial group living in relative poverty. (6)

Down

1. This area is beyond the main built-up area but has land use linked to the urban area. (5/5/6)
2. This person travels to and from work, from home, every day. (8)
4. This slum dwelling is a temporary or poorly-built construction, often on land that does not belong to the householder. (6)
6. This zone is on the edge of the CBD. It has been badly affected by change and may have derelict land and buildings, and high rates of crime and social problems. (8)

Urbanisation and the environment

Pollution

Air pollution

KEY IDEAS

→ Urban areas have less clean air than the surrounding countryside.

→ In MEDCs, strict regulation of vehicles and industrial plants has greatly reduced air pollution over the last 60 years.

→ It is in the major cities of LEDCs and NICs that the highest levels of air pollution occur.

Pollutant	Source	Problem
Carbon monoxide (CO)	• Vehicle exhausts	• Reduces the supply of oxygen to the heart
Carbon dioxide (CO_2)	• Vehicle exhausts • Power stations • Industrial processes • Domestic heating	• 'Greenhouse gas' which contributes to global warming
Nitrogen dioxide (NO_2) - one of a group of gases called nitrogen oxides	• Vehicle exhausts • Power stations	• Irritates the lungs • Leads to ozone formation
Ground level ozone (O_3)	• Reactions involving vehicle exhausts in the presence of sunlight	• Photochemical smog leading to irritations of the respiratory tract and eyes
Particulate matter	• Construction dust • Soot from open fires • Diesel vehicles	• Smog (fog + smoke) causes respiratory diseases
Sulfur dioxide (SO_2)	• Power stations • Vehicle exhausts	• Lung irritation • Acid rain
Hydrocarbons (including benzene)	• Vehicle exhausts	• Contributes to the formation of ground level ozone
Lead	• Exhaust gases from leaded petrol	• Harms the kidneys, liver, nervous system and other organs

Solutions:
- Laws to control the emissions from industry and housing, including smoke-free zones
- Carrying out checks on vehicle exhausts and removing polluting vehicles
- Higher taxes for the most polluting vehicles
- Introducing lead-free petrol
- Reducing the amount of electricity generated from thermal power stations
- Developing new power stations that do not release carbon dioxide into the air

Water pollution

KEY IDEAS

→ Raw sewage in rivers and groundwater.
→ Contamination of drinking water leading to health issues such as diarrhoea and dysentery.
→ This is a particular problem in some LEDCs.

Visual pollution

KEY IDEAS

→ Ugly buildings and industry   → Graffiti
→ Derelict land   → Litter

Solutions:
- Stricter planning regulations
- Improve refuse collection

Noise pollution

KEY IDEAS

→ Cars and lorries, trains, aircraft taking off and landing.
→ Factories, large congregations of people such as football crowds, noise in residential areas, for example from radios and parties.

Solutions:
- Laws which limit the noise from factories and homes.
- Separating noisy areas from residential areas.
- Building solid fences along motorways and major roads to reduce the noise reaching residents.

Urban sprawl

This is the spreading outwards of a city and its suburbs leading to changes in the surrounding rural area. It occurs particularly in countries like the USA, Canada and Australia where urban areas tend to have low-density suburbs.

Problems:
- High car dependence.
- Inadequate facilities within the spreading suburbs, such as entertainment, shops, doctors and transport.
- High cost of providing social facilities.

- High costs for public transport.
- Lost time and productivity for commuting.
- High levels of racial and socio-economic segregation.
- Changing character of the countryside and loss of the rural way of life. Rural areas and villages become urbanised and turned into **dormitory villages** for long distance commuters.

Solution:
- Green belts

Rural-urban fringe

This is the transition zone outside the main built-up area where urban and rural land uses are mixed. The rural-urban fringe will have agricultural land but with other types of land use linked to the urban area.

These include:
- roads, especially motorways and bypasses
- recycling facilities and landfill sites
- park and ride sites
- airports
- hospitals
- sewage facilities
- large out-of-town shopping facilities, such as large supermarkets
- golf courses
- parks and nature reserves.

Practice questions

1. The following table shows the quality of life in four large cities.

City	Persons per room	Percentage of houses with water and electricity	Infant mortality rate (per thousand births)	Noise index
Tokyo, Japan	0.9	100	5	4
New York, USA	0.5	99	10	8
Mexico City, Mexico	1.9	94	36	6
Kolkata, India	3.0	57	46	4

a) Compare the quality of life in the cities in MEDCs and the cities in LEDCs.
b) Suggest two other factors which affect quality of life. Justify your choice.

2. Study the following paragraph about Shanghai, a rapidly growing city in China.

Less than 60% of waste water and less than 40% of sewage flows are intercepted and treated. The Huangpu river receives 4 million cubic metres of untreated human waste every day. 30 000 tonnes of building waste are generated every day and landfill sites are almost full. Almost all households have access to piped water, electricity, and waste disposal. Some organic waste is now used as fertiliser in the surrounding rural area.

Coal-fired power stations provide 75% of China's electricity, leading to emissions of nitrogen oxides, particulate matter and sulfur dioxide. Emissions from motor vehicles are a problem in the strong sunlight. Some reductions in these emissions have been achieved. Strategies to improve air pollution include: upgrading the underground railway system, linking the airport to the city by a magnetic levitation train, and pedestrianisation.

a) Explain the health consequences of the water and air pollution in Shanghai.

b) Suggest other strategies which could reduce air pollution.

3. What is urban sprawl? For an example which you have studied, describe the causes of urban sprawl.

4. Which **one** statement in each of the following groups is **correct**?

a) Vehicles are a major source of air pollution by methane.

b) Vehicles are a major source of air pollution by nitrogen oxides.

c) Vehicles are a major source of air pollution by ozone.

d) Vehicles are a major source of air pollution by oxygen.

a) The greatest levels of air pollution are in rural areas of LEDCs.

b) The greatest levels of air pollution are in rural areas of MEDCs.

c) The greatest levels of air pollution are in urban areas of LEDCs.

d) The greatest levels of air pollution are in urban areas of MEDCs.

a) Air pollution by carbon dioxide is a major cause of cancer.

b) Air pollution by carbon dioxide is a major cause of respiratory disease.

c) Air pollution by carbon dioxide is a major cause of global warming.

d) Air pollution by carbon dioxide is a major cause of poisoning.

a) Air pollution by sulfur dioxide is mostly caused by power stations.

b) Air pollution by sulfur dioxide is mostly caused by vehicles.

c) Air pollution by sulfur dioxide is mostly caused by industry.

d) Air pollution by sulfur dioxide is mostly caused by domestic heating.

5. Which **one** statement in each of the following groups is **incorrect**?

a) Urban sprawl causes high levels of dependence on cars.
b) Urban sprawl causes lack of services in the suburbs.
c) Urban sprawl causes high costs of transport.
d) Urban sprawl causes water pollution.

a) Land use in the rural-urban fringe includes motorways and bypasses.
b) Land use in the rural-urban fringe includes heavy industry.
c) Land use in the rural-urban fringe includes landfill sites.
d) Land use in the rural-urban fringe includes park and ride sites.

a) Urban sprawl causes high levels of crime.
b) Urban sprawl causes loss of time due to commuting.
c) Urban sprawl causes racial and social segregation.
d) Urban sprawl causes the character of rural areas to change.

a) Land use in the rural-urban fringe includes airports.
b) Land use in the rural-urban fringe includes large supermarkets.
c) Land use in the rural-urban fringe includes government buildings.
d) Land use in the rural-urban fringe includes golf courses.

6. Match the terms with their definitions.

	Term		Definition
A	Smog	1	Cities that have grown outwards and have merged with other towns and cities to produce very large settlements.
B	Conurbations	2	A small settlement where many of the people commute to work in another settlement.
C	Dormitory village	3	A mixture of fog and pollutants.
D	Urban sprawl	4	Smog produced by a complex set of reactions of strong sunlight on nitrogen oxides and hydrocarbons from vehicle exhausts. It includes ozone.
E	Photochemical smog	5	The area just beyond the main built-up area which had distinctive land use linked to the urban area.
F	Rural-urban fringe	6	The movement of population from towns back to rural areas.
G	Counter-urbanisation	7	The process of large urban areas growing outwards rapidly into the surrounding rural area.

Theme 1: Population and settlement

Exam-style questions

1. a) Use information from the bar graphs to answer the following questions.

Population by continent, 2006 millions of people

Oceania 33

Europe 728	Asia 3958	Africa 916	North America 518	South America 376

Land area by continent thousands of square kilometres

Europe 10 498	Asia	Africa	Oceania	North America	South America	Antarctica

    i) Name the continents with
- a population too low to show on the graph,
- the largest land area. [1]

    ii) Name **one** continent that has a larger population than expected for its size. [1]

    iii) Without doing calculations, state which continent has the largest population density. [1]

    iv) Showing your working, calculate the average population density of Europe. [2]

    v) Most MEDCs are in Europe, North America and Oceania. Use the bar graphs to compare the population in those continents with the other continents which include mainly LEDCs. [2]

    vi) Is it possible to conclude from the evidence on the diagrams that a continent is under- or over-populated? Explain your answer. [2]

b) Suggest **four** physical reasons why some continents have lower populations than expected for their sizes. [4]

c) Describe problems for the government of a country that result from it being under-populated. [5]

d) Name a country you have studied which has densely populated areas. Describe the locations of the densely populated areas and explain why they have become densely populated. [7]

2. a) Study the diagrams which show information about population.

Population structure, 2006

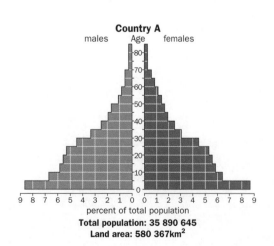

Country A

Total population: 35 890 645
Land area: 580 367km²

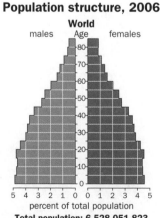

World

Total population: 6 528 051 823
Land area: 148 940 000km²

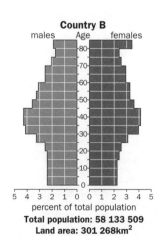

Country B

Total population: 58 133 509
Land area: 301 268km²

      i) Name the type of diagrams.   [1]

      ii) Suggest **two** possible reasons for the difference in
           the 0 to 4 and 5 to 9 age groups in country A.   [2]

      iii) Describe the population structure of the world.   [3]

      iv) Compare the population structure of country
           A with that of the world.   [3]

      v) Explain how the population structure of country B can be
           explained by birth rates, death rates and life expectancy.   [4]

b) Suggest the difficulties that result from having a population
structure like that of country B.   [5]

c) Name a country from which many people have migrated to another
country or countries. With reference to specific movements, explain
why so many people left the country.   [7]

3. a)

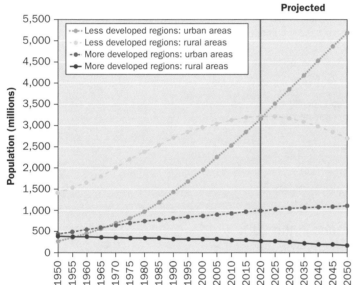

      i) Explain the meaning of the term *urbanisation*.   [1]

      ii) Study the graphs of population in rural and urban areas. In 2000,
          which of the four regions had the highest and the lowest populations?   [2]

      iii) Describe the changes in population shown by the graphs.   [3]

      iv) Explain the causes of the changes shown on the graphs.   [4]

b) The table shows the population of two rural areas in England and
the total population of England.

	Southwold (Suffolk)	Brancaster (Norfolk)	England
% aged 0–4	3.3	2.4	6.0
% aged 5–15	10.8	7.3	14.2
% aged 16–19	4.0	3.1	4.9
% aged 20–44	19.5	20.3	35.3
% aged 45–64	24.3	32.0	23.8
% aged over 65	38.1	34.9	15.8
Total population (2001)	4025	1484	49 000 000

▲ The age breakdown in Southwold and Brancaster, compared with England as a whole

i) Compare the rural populations with the population of England. [3]

ii) Explain the effects of population change in rural areas in MEDCs. [5]

c) For an example of a city you have studied, describe the problems caused by rapid urban growth. [7]

4. a) The diagram shows a model of urban land use for a city in an LEDC.

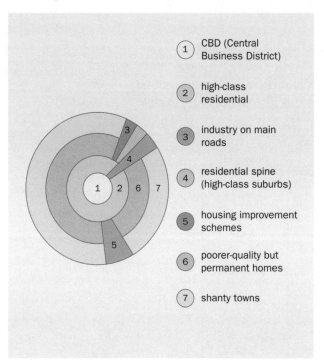

1. CBD (Central Business District)

2. high-class residential

3. industry on main roads

4. residential spine (high-class suburbs)

5. housing improvement schemes

6. poorer-quality but permanent homes

7. shanty towns

i) Name two types of land use found in the CBD. [1]

ii) Describe the distribution of industry in the model of urban land use. [2]

iii) Describe the distribution of housing types in the model of urban land use. [3]

iv) Give reasons for the distribution of housing types shown. [4]

b)

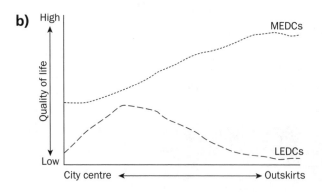

i) Using the diagram only, describe the differences in quality of life between LEDC cities and MEDC cities. [3]

ii) Give reasons for these differences. [5]

c) For a city you have studied, describe its morphology (distribution of land use types), explaining how well it fits models of urban land use. [7]

Plate tectonics

KEY IDEAS

→ The Earth's surface is made up of a series of sections known as **plates**.

→ The plates are, on average, about 50 kilometres thick and include the Earth's **crust** and the upper part of the layer below, which is called the **mantle**.

→ The plates are relatively cold and rigid. The rocks underneath the plates have temperatures of more than 1300°c and behave plastically.

→ The plates can move relative to each other, flowing over the hotter, more plastic rocks below, which act like a lubricated layer.

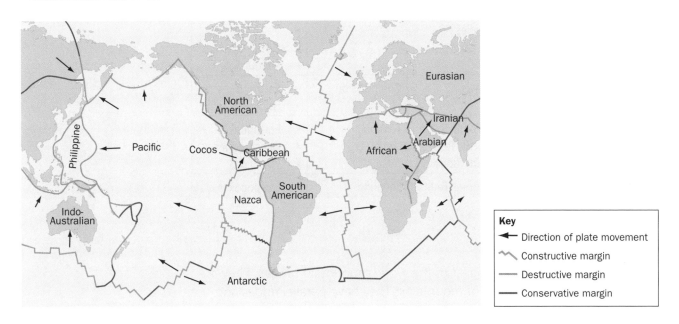

Key
← Direction of plate movement
⋏ Constructive margin
— Destructive margin
— Conservative margin

Why do the plates move?

Deep within the Earth heat is being produced by radioactivity. At the hotter areas the plastic rocks in the Earth's mantle become lighter and rise, causing convection currents. These convection currents drag the rigid plates above them causing them to move.

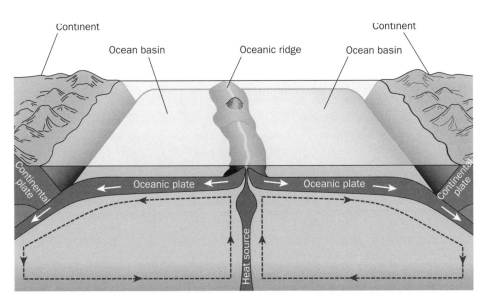

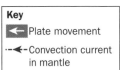

Key
← Plate movement
--◄- Convection current in mantle

Plate boundaries (plate margins)

Type of plate boundary	Examples	Type of stress	Features
Destructive with an oceanic plate and a continental plate	Andes	Compression	• Earthquakes • Fold mountains • Volcanoes • Ocean trenches • Subduction
Destructive with two oceanic plates	Japan Philippines West Indies	Compression	• Earthquakes • Island arcs • Volcanoes • Ocean trenches • Subduction
Destructive with two continental plates	Himalayas	Compression	• Earthquakes • Fold mountains
Constructive	Mid-Atlantic Ridge East Pacific Rise Carlsberg Ridge	Tension	• Earthquakes • Ocean ridges • Volcanoes
Conservative	San Andreas Fault	Shearing	• Earthquakes

Fold mountains

Fold mountains form the highest of the world's mountain ranges. They are long, relatively narrow belts of mountains. They have parallel ridges and valleys and the main range is made up of a series of ranges. Flatter areas form plateaux within the mountains. Active volcanoes form high conical mountains within the ranges.

Fold mountains have been formed where powerful **compression** has squeezed the layers of rocks.

Fold mountain ranges such as the Andes tend to be sparsely populated but communities are found in these areas. People live in high mountains and practise agriculture, have tourist industries and hydro-electric power.

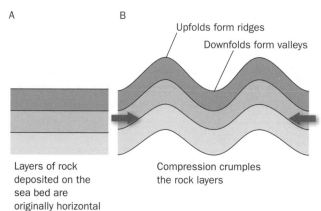

A   B   Upfolds form ridges

Downfolds form valleys

Layers of rock deposited on the sea bed are originally horizontal

Compression crumples the rock layers

Practice questions

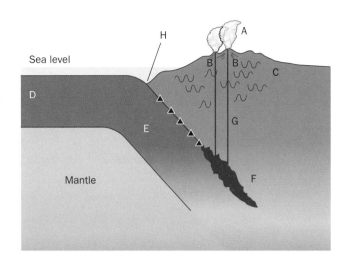

1. a) Why does a destructive margin have this name?

b) Name an example of a destructive margin where a continental plate meets an oceanic plate.

c) On the diagram draw arrows to show the directions of plate movement.

d) On the diagram, name the features labelled A to H.

e) On the diagram label the positions of shallow focus earthquakes and deep focus earthquakes.

Key

B	Volcano
▲	Earthquake focus

2.

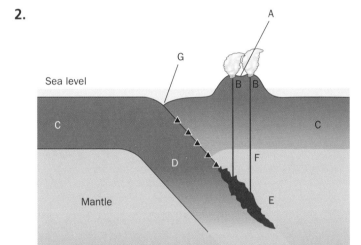

Key

B	Volcano
▲	Earthquake focus

a) Name an example of a destructive margin involving two oceanic plates.

b) On the diagram draw arrows to show the directions of plate movement.

c) On the diagram, name the features labelled A to G.

d) On the diagram label the positions of shallow focus earthquakes and deep focus earthquakes.

3. a) What type of plate margin is shown on the diagram?

b) Name an example of this type of plate margin.

c) Why do fold mountains and earthquakes occur here but not volcanoes?

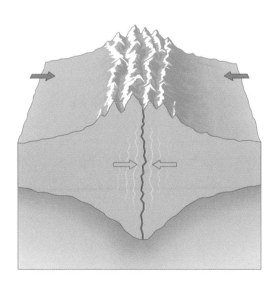

4. Name the types of stress shown below.

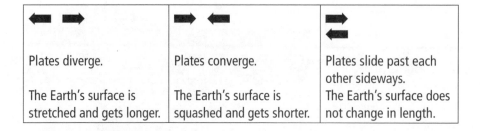

Plates diverge.	Plates converge.	Plates slide past each other sideways.
The Earth's surface is stretched and gets longer.	The Earth's surface is squashed and gets shorter.	The Earth's surface does not change in length.

5.

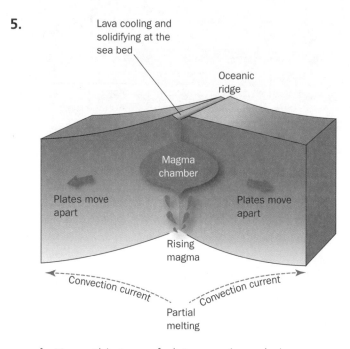

a) Name this type of plate margin and give an example.
b) Explain how new oceanic plate is created at this type of plate margin.
c) What will happen to an ocean which has this type of plate margin and no other plate margins?

6. List the advantages and disadvantages of living in fold mountains.

7. Complete the following crossword.

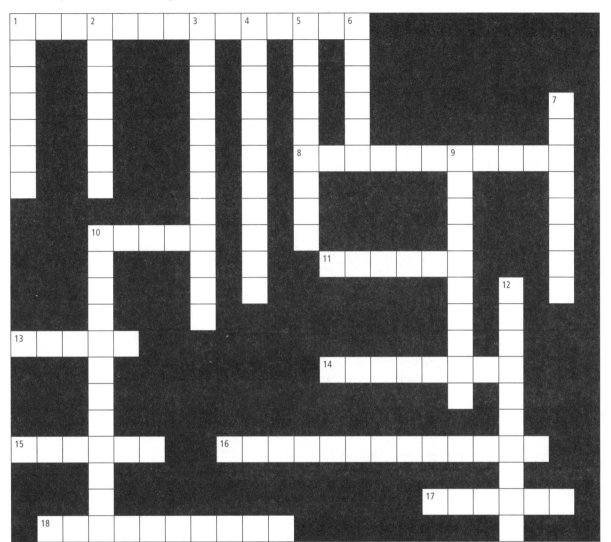

Across	Down
1. The concept that the Earth's surface is made up of a number of rigid sections that can slowly move over time. (5/9)	1. A solid which can flow and be deformed without fracturing. (7)
8. A plate margin where two plates converge and one is destroyed. (11)	2. Type of stress where forces pull in opposite directions causing stretching. (7)
10. The outer layer of the Earth, between 6 and 90km thick. (5)	3. A plate margin where plates slide past each other and plate is neither created nor destroyed. (12)
11. The middle layer of the Earth between the crust and the core. (6)	4. A long, narrow area of the ocean floor about 10km deep, usually at a destructive plate margin at the edge of an ocean. (5/6)
13. The planet of the solar system which is the third nearest to the Sun. (5)	5. A chain of islands forming a crescent shape, formed from volcanoes at a destructive margin involving two oceanic plates. (9)
14. Where rocks have been crumpled in a downwards direction like a letter U. (8)	6. The force per unit area acting on an object. (6)
15. The upper, colder, rigid parts of the Earth's surface, made up of the crust and upper mantle. (6)	7. Type of stress where forces cut across each other. (8)
16. The process of the breakdown of the nuclei of unstable atoms, releasing heat and radioactive energy. (13)	9. Type of current in liquids and gases caused by heating. (10)
17. Where rocks have been crumpled in an upwards direction like an upside down letter U. (6)	10. A plate margin where plates diverge and new oceanic plate is created. (12)
18. A broad, high belt of the ocean floor which is much higher than the surrounding area. Formed by volcanic activity at constructive plate margins. (5/5)	12. The process where plates converge and one plate is forced beneath the other. (10)

Volcanoes and earthquakes

Volcanoes

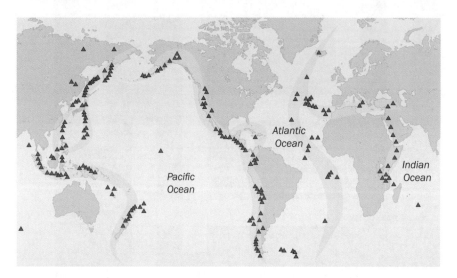

Key

▲ Volcano

Volcano belts

What comes out of a volcano?

Gases

You should be able to name some of the gases and explain how they control eruptions.

Liquids

Lava is molten rock material erupted onto the Earth's surface as lava flows.

Solids

These are known as pyroclastic material.

Types of volcano

	Shield Volcano	Stratovolcano
Slope angles	Gentle	Steep
Plate tectonic setting	Constructive margins Mid-plate volcanoes	Destructive margins with an oceanic plate
Products	Mostly lava	Lava and pyroclastics (ash)
Lava viscosity	Non-viscous	Viscous
Type of eruption	Continuous and non-violent	Explosive with dormant phases

You should know about the following other features of volcanoes:

- Craters
- Calderas
- Parasitic cones
- Lava domes

Earthquakes

Earthquakes are caused by plate movements, either towards each other, away from each other or sliding past each other. The plates do not always move at a constant rate. They are often 'stuck' in one position. Stress builds up as the plates try to move. There is then a sudden movement (along a crack in the Earth called a **fault**). Energy is released and vibrations travel through the Earth as an earthquake wave or seismic wave.

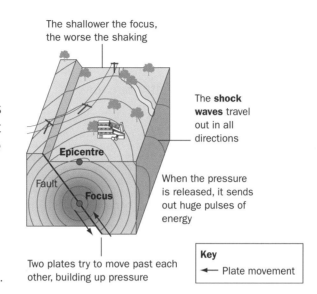

The shallower the focus, the worse the shaking

The **shock waves** travel out in all directions

Epicentre

Fault

Focus

When the pressure is released, it sends out huge pulses of energy

Two plates try to move past each other, building up pressure

Key
← Plate movement

KEY IDEAS

→ The point within the Earth where the earthquake originates is called the **focus**. The point on the Earth's surface directly above the focus is called the **epicentre**.

→ The effects of an earthquake are described on a 12 point scale called the **Mercalli Scale** of earthquake intensity.

→ Earthquakes have a single figure on the **Richter Scale** of magnitude which measures the total amount of energy released by an earthquake.

→ Lines of equal intensity drawn on a map around the epicentre of an earthquake are called **isoseismal lines**.

The amount of damage an earthquakes causes will be affected by the following factors:

- Energy released (Richter Scale).
- Depth of the focus beneath the surface (shallower earthquakes have a greater effect).
- Density of population.
- Whether or not the buildings have been built to withstand earthquakes.
- How solid the bedrock is; weak sands and clays can turn to liquid, known as liquefaction, causing buildings to collapse.
- Ability of an area to recover from a major earthquake is affected by how wealthy a country is.

Practice questions

1. **a)** Name the type of volcano shown in the diagram and name an example.

   b) At what type of plate margins do these volcanoes form?

   c) Describe the size and shape of these volcanoes and how this is linked to the type of lava.

   d) Describe how these volcanoes erupt and how the type of eruption is linked to the type of lava.

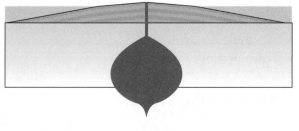

2. **a)** Name the type of volcano shown in the diagram and name an example.

   b) At what type of plate margins do these volcanoes form?

   c) Describe the shape of these volcanoes and how this is linked to the type of lava.

   d) Describe how these volcanoes erupt and how the type of eruption is linked to the type of lava.

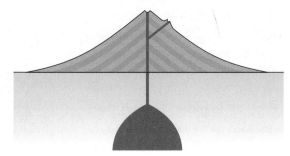

3.

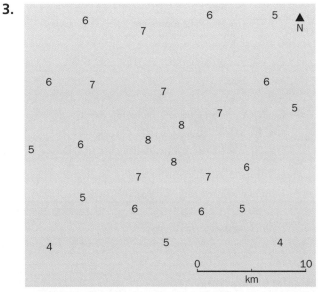

a) The map shows the effects of an earthquake on the Mercalli Scale. Draw isoseismal lines on the map.

b) Mark the position of the epicentre on the map.

c) Describe the effects of the earthquake at the epicentre.

d) The earthquake was reported to have a magnitude of 6.5. Why are some of the numbers on the map bigger than this?

Value	Intensity	Description
1	Instrumental	Not normally felt. Animals uneasy.
2	Feeble	Felt only by a few people at rest.
3	Slight	Vibrations like a lorry passing. Felt by people at rest.
4	Moderate	Felt indoors by many. Cars rock.
5	Rather strong	Sleepers awakened. Some windows broken.
6	Strong	Bells ring. Trees sway. Loose objects fall.
7	Very Strong	Difficult to stand up. People run outdoors. Walls crack.
8	Destructive	Collapse of some buildings. Trees fall.
9	Ruinous	Ground cracks. Pipes break.
10	Disastrous	Landslides. Many buildings destroyed.
11	Very disastrous	Few buildings left standing.
12	Catastrophic	Total damage. Ground surface rises and falls in waves. Objects thrown into the air.

▲ The Mercalli Scale

4. Match the terms with their definitions.

	Term		Definition
A	Tsunami	1	Fine dust produced by a volcano. The finest pyroclastic material.
B	Crater	2	An ocean wave produced when there is movement of the sea bed.
C	Earthquake	3	A volcano which has not erupted for some time but may erupt in the future.
D	Shield volcano	4	Using scientific instruments on satellites above the Earth to provide information about the Earth.
E	Extinct	5	Molten rock on the Earth's surface.
F	Mercalli Scale	6	The point on the Earth's surface directly above an earthquake focus.
G	Focus	7	A volcano which will never erupt again.
H	Geothermal power	8	A scale measuring the total amount of energy released by an earthquake.
I	Richter Scale	9	The point in the Earth where an earthquake occurs.
J	Magma	10	A volcano made from alternate layers of lava and ash.
K	Lahar	11	A mudflow which contains material from volcanic eruptions.
L	Shearing	12	When a volcano erupts sideways with great force producing gas and pyroclastic material.
M	Lava	13	A crack in the rocks of the Earth's crust where the rocks move and are displaced.
N	Lateral blast	14	Molten rock below the Earth's surface.
O	Magma chamber	15	A huge chamber, several kilometres across which stores molten rock beneath a volcano.
P	Remote sensing	16	A scale which describes the effects of an earthquake.
Q	Mudflow	17	A powerful current of water mixed with mud, rocks and boulders running down a river channel, valley or slope.
R	Pyroclastic	18	Refers to the solid material produced during a violent volcanic eruption.
S	Dormant	19	The depression at the top of a volcano.
T	Ash	20	Using the heat contained deep in the Earth which is more concentrated in volcanic areas.
U	Fault	21	Type of stress where forces cut across each other.
V	Stratovolcano	22	A large volcano with gentle sides, built up by non-viscous lava and very little ash.
W	Epicentre	23	Violent shaking of the upper layers of the Earth when energy is released as the rocks crack and move.

12 Weathering

Weathering is the decay and disintegration of rocks *in situ*, involving physical, chemical and biological processes, resulting from the conditions in the atmosphere. This means that the rocks that form the Earth's surface are slowly broken down over long periods of time.

Weathering is often confused with **erosion**. Weathering is different because it takes place *in situ*, in other words without movement. The processes that carry out weathering do not transport the products away. Rivers and waves in the sea (and the wind and glaciers) carry out erosion, they do not carry out weathering.

Type of weathering	Processes	Features formed
Physical (mechanical)	Freeze-thaw action	Scree
	Exfoliation	Exfoliation domes
Chemical	Carbonation	Deep cracks in rock, caves
	Oxidation	Rusty red or orange staining
Biological	Wedging effect of tree roots Release of acids and carbon dioxide Vegetation trapping water	

Freeze-thaw action

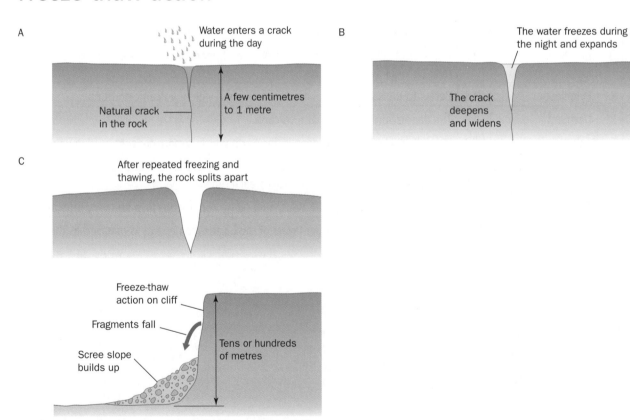

A — Water enters a crack during the day
Natural crack in the rock
A few centimetres to 1 metre

B — The water freezes during the night and expands
The crack deepens and widens

C — After repeated freezing and thawing, the rock splits apart

Freeze-thaw action on cliff
Fragments fall
Scree slope builds up
Tens or hundreds of metres

Exfoliation

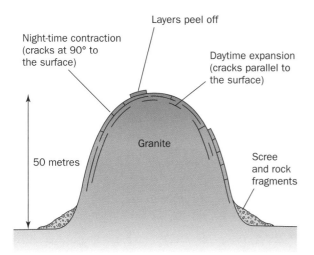

Layers peel off

Night-time contraction
(cracks at 90° to
the surface)

Daytime expansion
(cracks parallel to
the surface)

Granite

50 metres

Scree
and rock
fragments

Carbonation

Naturally occurring carbon dioxide (CO_2) in the air is dissolved in rainwater (H_2O) to produce a weak acid, **carbonic acid** ($H_2O + CO_2 \rightarrow H_2CO_3$).

This weak acid dissolves the calcium carbonate in limestone to form a solution which is washed away as a solution of calcium ions (Ca^{2+}) and hydrogen carbonate ions (HCO_3^-). Any impurities in the limestone are left behind as clay or sand.

Oxidation

Oxidation is the addition of oxygen to a mineral. The process affects rocks in dry areas where there is oxygen in contact with the rocks (oxic conditions).

Rocks that contain iron are affected by oxidation.

Wedging effect of tree roots

Where the soil is shallow, the seeds and roots of trees find their way into natural cracks in the bedrock. As the seeds germinate and the roots get bigger, they have the effect of making the cracks wider and deeper, eventually breaking up the bedrock.

Release of acids and carbon dioxide

Decaying plant matter, for example on the floor of a forest or beneath a cover of grass, produces chemicals such as **humic acids** and carbon dioxide. These chemicals then cause chemical weathering processes to occur.

Vegetation trapping water

Beneath a cover of vegetation water may be trapped, speeding up the rates of chemical weathering.

Factors affecting weathering

Rock type	Presence of lines of weakness	• Allow water to penetrate the rock and increase both physical and chemical effects of weathering. These weaknesses also control the size and shape of the weathered fragments.
	Grain size	• In general the bigger the grains or crystals the faster the rate of weathering. • Crystalline rocks have a greater resistance than rocks made out of grains or fragments.
	Mineral composition	• Limestone is made from calcium carbonate and is susceptible to carbonation. • Rocks which contain iron minerals are prone to oxidation. • One of the most common minerals, quartz, is chemically resistant and does not weather chemically.
Climate	Frost climates	• Freeze thaw action. • Chemical weathering is slow.
	Temperate areas	• Freeze-thaw action is not important and exfoliation does not occur. The main type of weathering is chemical.
	Deserts	• Exfoliation. • Rates of weathering are the slowest on Earth.
	Humid (wet) tropical areas	• Very rapid rates of chemical weathering.

Practice questions

1.

	Polar region	Humid temperate region	Tropical desert region	Tropical rainforest region
Temperature	Low	Moderate	High	High
Rainfall	Low	Moderate	Low	High
Depth of weathering				

Ground surface

30 metres

Unweathered rock

Weathered rock

▲ The rate of weathering in different climates

a) Which processes occur in each of the climatic zones shown on the diagram?

b) Describe and give reasons for the differences in depth of weathering shown on the diagram.

2. Complete the following crossword.

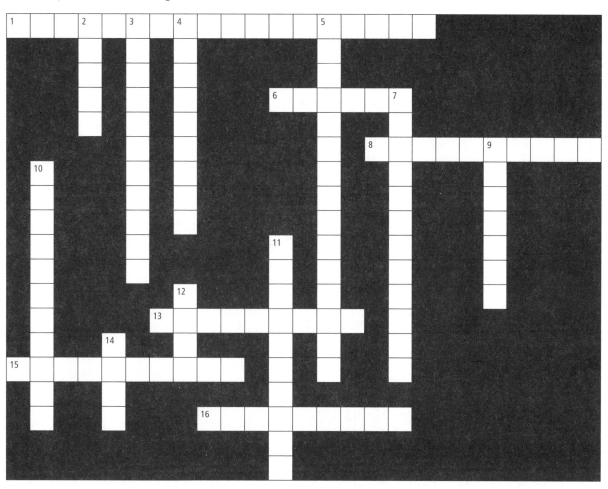

Across	Down
1. The breakdown of rocks at the Earth's surface involving changes in temperature. (8/10)	2. Angular rock fragments. (5)
6. Conditions which are waterlogged and there is little oxygen. (6)	3. Process of chemical weathering where limestone is attacked by carbonic acid produced in rainwater. (11)
8. Process of physical weathering occurring in cold climates involving water freezing in cracks causing the rocks to shatter. (6/4)	4. Common rock type containing a large amount of calcium carbonate. (9)
13. The addition of oxygen to a mineral (or the removal of an electron from an atom or ion). (9)	5. Large, rounded hill of bare rock occurring in a desert. (11/4)
15. Type of weathering when plant or animal matter plays an important part. (10)	7. This chemical makes the rainfall naturally acidic. (8/4)
16. Chemical released when plant (or animal) matter decays. (5/4)	9. The wearing away and removal of rock and weathered debris by running water and the wind. (7)
	10. Weathering when layers of rock peel off parallel to the surface. (12)
	11. The decay and disintegration of rocks in situ, involving physical, chemical and biological processes. (10)
	12. Conditions when oxygen is freely available for chemical reactions. (4)
	14. The thin layer at the Earth's surface formed from minerals produced by weathering and decaying organic matter. (4)

River processes (1)

The upper course of the river

Erosion

There are four processes of river erosion.

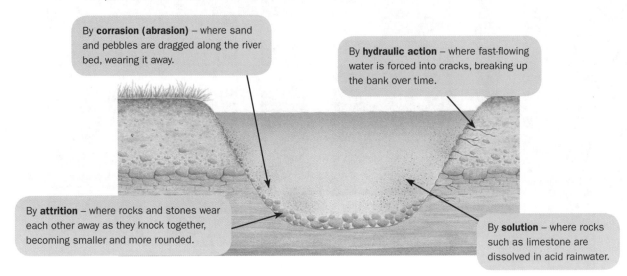

By **corrasion (abrasion)** – where sand and pebbles are dragged along the river bed, wearing it away.

By **hydraulic action** – where fast-flowing water is forced into cracks, breaking up the bank over time.

By **attrition** – where rocks and stones wear each other away as they knock together, becoming smaller and more rounded.

By **solution** – where rocks such as limestone are dissolved in acid rainwater.

Transport

There are four processes of river transport.

The boulders, pebbles, sand, silt and mud being transported are called the river's **load**.

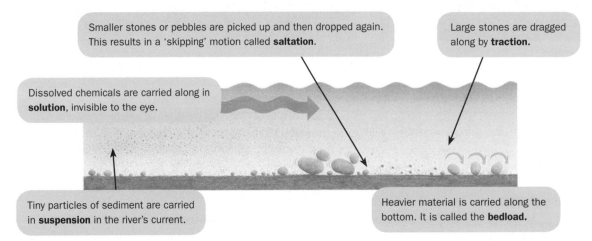

Smaller stones or pebbles are picked up and then dropped again. This results in a 'skipping' motion called **saltation**.

Large stones are dragged along by **traction.**

Dissolved chemicals are carried along in **solution**, invisible to the eye.

Tiny particles of sediment are carried in **suspension** in the river's current.

Heavier material is carried along the bottom. It is called the **bedload.**

Deposition

Deposition occurs when a river loses velocity (energy). This can be caused by:

- a decrease in gradient
- a decrease in river flow (discharge) as water drains away after heavy rain
- the river meeting the sea or a lake, often forming a delta
- the river flowing more slowly on the inside of bends.

Discharge

This is the volume of water flowing down the river at any time. It is measured in cubic metres per second, often referred to as **cumecs**. In climates with wet and dry seasons, or which are affected by melting snow in spring, the discharge can vary considerably.

River channel patterns

The river channel seen on a map or from the air has three main forms as shown by the following diagrams.

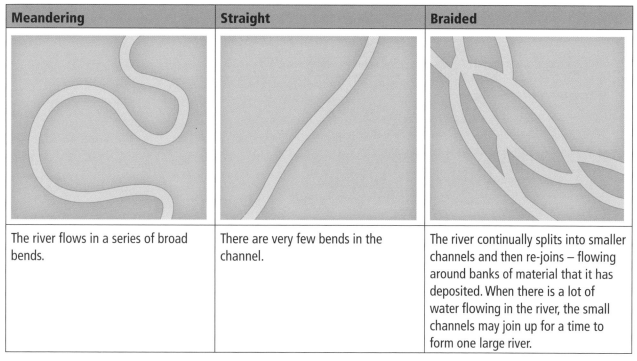

Meandering	Straight	Braided
The river flows in a series of broad bends.	There are very few bends in the channel.	The river continually splits into smaller channels and then re-joins – flowing around banks of material that it has deposited. When there is a lot of water flowing in the river, the small channels may join up for a time to form one large river.

▲   Different river channel patterns

Landforms of the upper course of a river

Shape of the valley	The place where a river starts its course is known as the **source**. The shape of the valley in cross section is a V-shape. The valley floor is very narrow and the river channel may occupy the whole of it. Large boulders in the river – the bed load – are only moved after heavy rainfall when the river becomes a powerful torrent. The river is carrying out **vertical erosion**. The gradient of the river may be quite steep.
Potholes	Smooth, rounded, hollows in the bedrock of the bed of the channel. They are often about 30cm across. They are formed by stones being trapped in hollows on the bed of the river. Eddies in the water swirl the stones around causing **corrasion**. The hollows become deeper and wider and eventually join together.
Rapids	Places where the water is shallow and the bed of the river is rocky and irregular. The gradient is often steeper than at other points in the river's course. Rapids are often a barrier to navigation on rivers. Rapids may be caused by a band, or bands, of hard rock in the river bed.
Waterfalls and gorges	Formed when a horizontal layer of hard rock lies over a layer of soft rock. The soft rock is eroded more quickly. Gradually a **plunge pool** develops. The splashing water and eddy currents in the plunge pool undermine the hard rock. The hard rock layer is left unsupported and eventually collapses. If the processes of undercutting and collapse are repeated over a long period of time, the waterfall will retreat upstream forming a deep, steep-sided valley called a **gorge**.
Interlocking spurs	In the upper course of a river where the valley is narrow, there are spurs of land on either side of the valley and the river winds around them.

Practice questions

1.

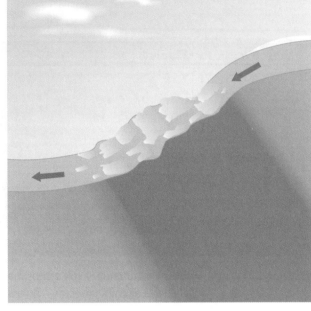

Copy the diagram and add the following labels:

- Hard rock
- Soft rock
- Rapids

2.

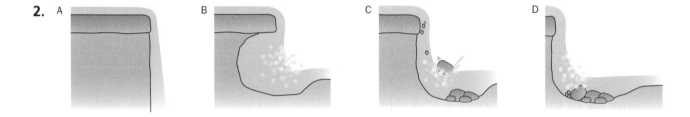

Copy the diagrams and add the following labels:

- Waterfall
- Gorge
- Plunge pool
- Hard rock
- Soft rock
- Overhang
- Original position of waterfall

3. Draw a labelled diagram of the upper course of a river to show steep sides, narrow V-shaped cross section and interlocking spurs.

4. Name the odd one out in each of the following lists:

   a) Corrosion, corrasion, attrition, saltation
   b) Hydraulic action, traction, suspension, solution
   c) Carbonation, oxidation, exfoliation, corrosion
   d) Energy, velocity, discharge, load
   e) Meandering, V-shaped, braided, straight

5. Choose an example of a river that you have studied. Describe the landforms in the upper course of the river and explain how they have formed.

6. Match the terms with their definitions.

	Term		Definition
A	Tributary	1	Process of erosion by which the force of running water alone removes material from the bed and banks.
B	Confluence	2	When the river loses energy and the load it is carrying is dropped.
C	Traction	3	Process of transport where mud and silt is held within the body of the water making it discoloured.
D	Corrosion	4	A smaller river which joins a larger river.
E	Potholes	5	Process of erosion where certain minerals in the rocks of the channel are dissolved by the river water.
F	Discharge	6	Rounded hollows in a rocky river bed.
G	Attrition	7	The volume of water passing down the river in a period of time. Usually measured in cubic metres per second.
H	Gorge	8	Process of erosion when fragments of the river's load become smaller and more rounded by collision with the bed and banks and each other.
I	Hydraulic action	9	Process by which a river (or waves in the sea, or ice or wind) wears away and removes rocks and sediment.
J	Load	10	A deep, steep-sided valley.
K	Plunge pool	11	The point where a river and a tributary meet.
L	Transport	12	Process of erosion when the bed and banks of the river are eroded by the abrasive effect of the river's load.
M	Rapids	13	The boulders, pebbles, sand, silt, mud and minerals in solution carried by a river.
N	Waterfall	14	The deep hollow in a river bed at the foot of a waterfall.
O	Saltation	15	The speed of flow of the river water downstream. Usually measured in metres per second.
P	Interlocking spur	16	The movement downstream of the river's load.
Q	Corrasion	17	Process of transport where material moves downstream in a series of hops.
R	Suspension	18	Ridges or shoulders of land which slope down towards a river from either side of the valley and which the river winds around.
S	Erosion	19	Process of transport where material rolls downstream staying in contact with the river bed.
T	Deposition	20	Where the river flows over a vertical drop in its long profile.
U	Velocity	21	An irregular, often steeper, section of the long profile of a river where the water is white.

14 River processes (2)

The middle and lower courses of a river

	Upper course	Middle course	Lower course
	(long profile diagram)		
Long profile	• Steep long profile • Rapids • Waterfalls • Lakes	• Profile becomes more gentle	• Gentle long profile
	(cross profile diagram)	*(cross profile diagram)*	*(cross profile diagram)*
Cross profile	• Cross profile steep and V-shaped • Valley floor narrow or non-existent	• Cross profile more gentle • Flood plain beginning to develop • Cross profile is often asymmetrical, with river cliffs and slip-off slopes	• Cross profile is very gentle • Wide flood plain

▲ The long and cross profiles of a river valley

In the middle course of the river vertical erosion becomes less important and **lateral** (sideways) **erosion** and deposition start to take over. In the lower course, vertical erosion may stop.

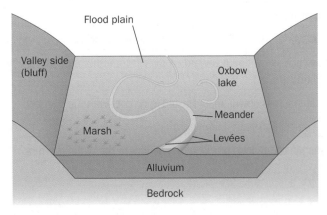

▲ Features of the lower course

Meanders and oxbow lakes	• Erosion on outer bends where water flows fastest – forms a **river cliff**. • Deposition on inside bends where water flows slowest – forms a **slip off slope**. • This increases the size of meanders until a narrow neck forms. • River breaks through during flooding. • Ends of oxbow lake sealed by deposition.
Flood plains	• Deposition of point bars on the insides of meanders. These deposits are spread across the valley as the meanders migrate. • Deposition of gravel on the river bed (part of the bed load). • Deposition of fine silt and mud (part of the suspension load) on the flood plain itself during floods when the river overflows its banks.
Levées	• Levées are formed when the river floods. As the water overflows the channel it slows down (loses energy) therefore more of the load, and a coarser part of the load, is deposited close to the channel making the banks higher. • During normal flows, the river deposits on the river bed within its channel, making the bed higher and causing the river bed to be higher than the surrounding flood plain.
Deltas	• Deltas form because the river carrying its load meets the still water of the sea or lake. The loss of velocity (energy) leads to deposition which builds up to form the delta. • Over time, the delta is gradually built out into the sea. • Deposition blocks the channels, which leads to the formation of distributaries.

Practice questions

1.

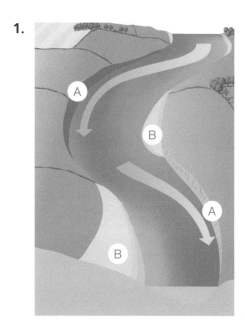

2.

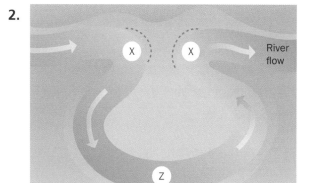

a) Name the feature of the river at Z.
b) What feature will form at Z in the future?
c) Name the process which will occur at X.

a) Describe the speed of the water at A.
b) What process occurs at A?
c) Name the landform which occurs at A.
d) Describe the speed of the water at B.
e) What process occurs at B?
f) Name the landform which occurs at B.

3. a) Name the landform shown on the diagram.

b) Name the process which occurs at X and Y.

c) When does this process occur at X and when does this process occur at Y?

d) What happened to the bed and banks of the river over a period of time?

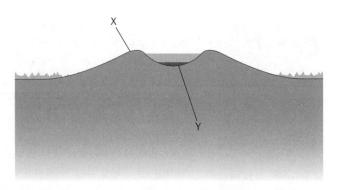

4.

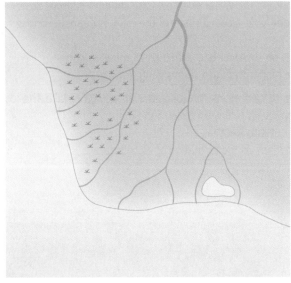

5. Name the odd one out in each of the following lists:

a) Flood plain, levée, oxbow, interlocking spurs

b) Delta, rapids, waterfalls, potholes

c) Gentle long profile, wide flood plain, deposition, vertical erosion

On the diagram label the following features:

- Sea or lake
- Lagoon
- Swamp or marsh
- Distributary
- Flat land

6. Match the terms with their definitions.

	Term		Definition
A	Braided channel	1	A bend in a river.
B	Levée	2	The area of flat land either side of a river which is occasionally flooded. Formed from deposits by the river.
C	Oxbow lake	3	The channels formed when a river divides, as in the braided channels in a delta.
D	Distributary	4	Raised banks on either side of a river flowing across a flood plain. They may cause the river to flow above the level of the flood plain.
E	Flood plain	5	The boulders, pebbles, sand, silt, mud and minerals in solution carried by a river.
F	Delta	6	Area of flat land where a river meets a sea or lake, formed by river deposits.
G	Meander	7	A channel which splits and re-joins.
H	Load	8	An old meander which has become cut off from the river. Often filled by vegetation.

7. Find the words which match the following descriptions. They may be shown horizontally or vertically.

W	X	C	O	X	B	O	W	U	B	X	I	O	P	N
J	K	L	I	E	B	N	S	C	Q	W	D	S	D	M
F	G	B	J	K	L	O	P	W	V	B	I	N	H	E
Q	S	R	X	C	V	B	N	E	Y	T	S	M	J	A
O	I	A	P	H	J	K	L	W	S	D	T	G	H	N
C	B	I	V	N	D	E	L	T	A	T	R	J	K	D
L	W	D	C	V	B	M	G	H	F	K	I	O	L	E
O	D	E	X	Z	C	V	N	M	W	I	B	P	E	R
A	S	D	F	G	H	K	L	O	P	T	U	Y	R	W
D	S	A	F	H	J	C	B	V	N	M	T	L	K	J
H	G	F	D	S	A	P	O	I	U	Y	A	T	R	E
W	Q	M	N	B	V	C	X	Z	L	K	R	H	G	F
S	L	E	V	E	E	Q	W	E	R	T	Y	I	U	O
Q	W	E	R	H	J	T	Y	U	I	O	P	L	K	J
F	L	O	O	D	P	L	A	I	N	L	K	D	A	S

a) This river channel splits and re-joins.
b) An area of flat land where a river meets a sea or lake.
c) The channels formed when a river divides.
d) The area of flat land either side of a river.
e) Raised banks on either side of a river flowing across a flood plain.
f) The material carried by a river.
g) A bend in a river.
h) An old meander which has become cut off from the river.

15 Marine erosion

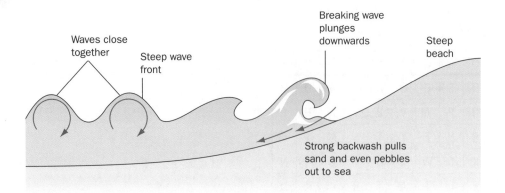

Waves close together

Steep wave front

Breaking wave plunges downwards

Steep beach

Strong backwash pulls sand and even pebbles out to sea

▲ Destructive waves

With destructive waves the backwash is stronger than the swash so beach material is pulled down the beach.

The four processes of marine erosion are the same as those for rivers.

The longer the load of eroded material is moved by the waves, the greater the effects of attrition; an angular boulder can be broken down to round pieces of shingle and eventually to small, round grains of sand.

Landforms formed by marine erosion

Cliffs	Cliffs are vertical or very steeply sloping rocks. Their slope depends on the nature of the rocks and the degree of wave attack at their bases. Cliffs in horizontal rocks are vertical. If the rock layers dip steeply away from the sea the cliff profile slopes more gently. Many cliffs have an indentation called a wave-cut notch at about the high tide level where wave attack has undercut them.
Wave-cut platforms	These solid rock platforms between the cliff base at high water mark and the low water mark slope gently towards the sea and may have rock pools. They are often covered with debris eroded from the cliffs.
Headlands	Headlands are areas of resistant rock projecting out to sea. They have cliffs along their sides.
Bays	Bays are approximately semi-circular shapes of sea extending in to the land with wide, open entrances from the sea. The land behind them is lower and more gently sloping than the headlands to either side of them.
Caves	These are indentations at the base of cliffs, formed where lines of weakness have been enlarged by erosion.
Arches	Arches are holes right through headlands and are formed by the joining of caves eroded into both sides of the headlands along a line of weakness.
Stacks	These vertical pillars of rock are isolated in the sea off headlands (to which they used to be attached.) They have flat tops if the rocks are horizontal. Others have pointed tops.

Cliffs and wave-cut platforms

Wave attack at the base of a cliff leaves the rock above unsupported. Eventually, it collapses. As this continues, the cliff is worn back, leaving a wave-cut platform where the cliff once stood.

Headlands and bays

Headlands and bays form on discordant coasts where the rocks lie at right angles to the sea and are subjected to differential erosion.

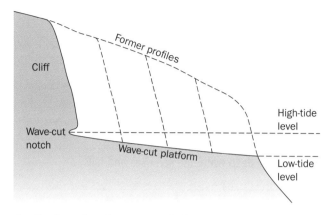

▲ The formation of a wave-cut platform

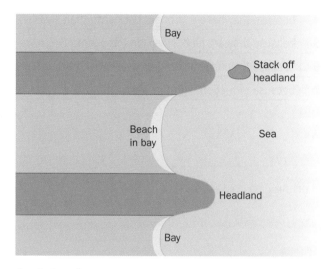

The less resistant rocks are more easily eroded than the resistant rocks, so they are worn back more quickly to form bays. The harder rocks resist erosion, forming headlands which protrude out into the sea between the bays.

▲ A discordant coast

Caves, arches and stacks

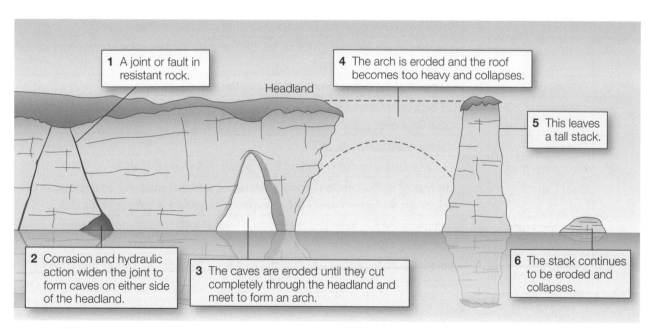

▲ How caves, arches and stacks form

Practice questions

1. Match the beginnings and endings to make explanations. (You will not use all the endings)

	Beginnings		Endings
A	When the wind blows over the sea it makes waves…	1	because the water particles cannot move in a circular manner.
B	In deep water the sea rises and falls…	2	because they have been driven by strong wind over a long distance of sea (the fetch).
C	The wave breaks in shallow water…	3	because individual water particles are moving in a circular manner.
D	Some waves are strong…	4	because they have a stronger backwash than swash.
E	Strong waves are destructive…	5	because they are high, steep, and frequent.
F	Destructive waves are called plunging breakers…	6	because they have a stronger swash than backwash.
		7	because of friction with the wind.

2. Which marine erosion process fits each statement?
   a) This type of marine erosion only erodes the load.
   b) These types erode the cliff, not the load.
   c) This type can erode both the cliff and the load in certain circumstances.
   d) This type cannot act without the load.

3. Match the terms with their definitions.

	Term		Definition
A	Sand	1	The forward movement of a wave up the shore.
B	Shingle	2	The area between the level of the lowest tides and the point reached by the highest storm waves.
C	Swash	3	Round beach material that is larger than sand but smaller than a boulder.
D	Shore	4	A process of erosion that dissolves soluble minerals, such as calcium carbonate in rock like limestone and chalk.
E	Coastline	5	The mean of the highest levels reached by the sea at high tides.
F	Corrosion	6	The high water mark on a lowland coast and the foot of cliffs on a steeply sloping coast. On a map, the line that separates land and sea.
G	Corrasion	7	A steep or vertical slope at the coast.
H	High water mark	8	A small rounded particle resulting from a long period of attrition.
I	(Sea) Cliff	9	This type of marine erosion occurs when waves trap and compress air in cracks in the rock. When the wave retreats the air expands violently and enlarges the crack.
J	Hydraulic action	10	An indentation bigger than a notch at the base of a cliff where rock has been removed by erosion.
K	Cave	11	A marine erosion process whereby the load is thrown against the cliff and chips bits of rock off it.

4. Complete the crossword.

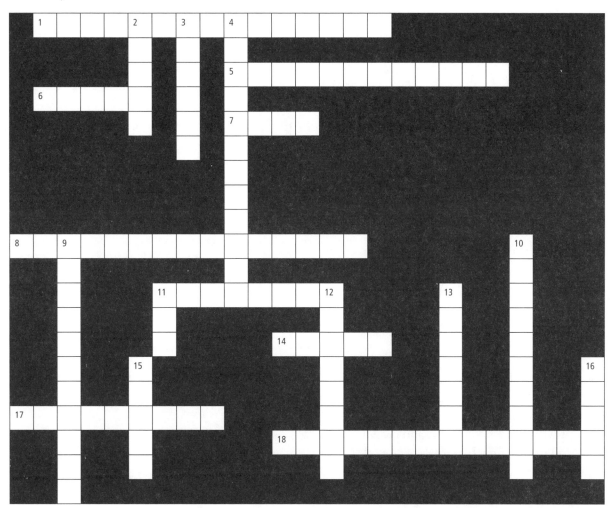

Across	Down
1. The feature is found at the foot of cliffs and its name tells you how it was formed. (7/8)	2. The highest part of a wave. (5)
5. An indentation near high water mark on a cliff. Its name tells you how it was formed. (7/5)	3. The lowest part of a wave. (6)
6. The eventual joining up of two of these by erosion on both sides of a headland forms a hole through it. (5)	4. This is the lower limit of wave erosion. (3/5/4)
7. A hole with a roof that extends all the way through a headland. (4)	9. This type of wave erodes. (11)
8. This type of marine erosion involves compression and expansion of air inside cracks in the rocks. (9/6)	10. This term is used for coastlines where the rocks are at right angles to the coast. (10)
11. The water in a wave that moves down the shore. (8)	11. A semi-circular area of water between two projections on a discordant coast. (3)
14. You would probably have to swim to reach this upstanding pillar of rock. (5)	12. A narrow, steep-sided piece of land which projects into the sea. (8)
17. This type of marine erosion makes the load rounder and smaller. (9)	13. This eroded particle is smaller than a boulder but bigger than a grain of sand. (7)
18. All headlands with stacks must have had at least one of these. (4/2/8)	15. The length of uninterrupted sea over which the wind blows to make a wave. (5)
	16. This part of the wave moves up the beach. (5)

Marine transportation and deposition

Marine transportation

Processes of marine transportation are the same as for rivers.

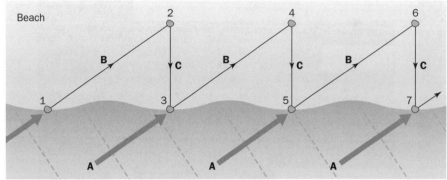

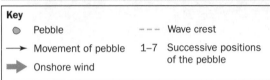

A Oblique onshore winds drive the wave crest at an angle to the shore.

B The swash of the wave carries the pebble at an oblique angle up the beach.

C The backwash of the wave brings the pebble straight down the beach under the influence of gravity. As this process is repeated, the pebble is moved along the beach. The direction of longshore drift is from left to right in this example.

Key

⬤	Pebble	– – –	Wave crest
→	Movement of pebble	1–7	Successive positions of the pebble
➡	Onshore wind		

▲ The process of longshore drift

Beach loss concerns local authorities because it makes the beach less attractive for tourists and removes the protection from erosion that the beach provides for the cliffs behind (wave energy is used on the beach). Groynes placed across the beach reduce longshore drift by trapping sand but this reduces the supply of sand for beaches further along the coast.

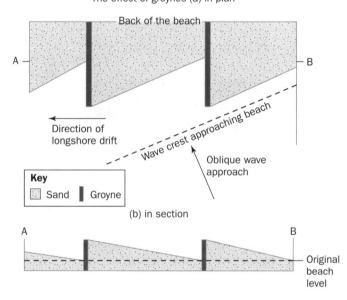

The effect of groynes (a) in plan

Key

▦ Sand	▮ Groyne

(b) in section

▲ The effect of groynes in plan and in section

Marine deposition

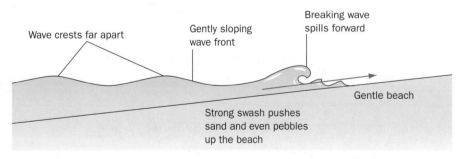

▲ Constructive waves

Constructive waves build up the beach because their swash is stronger than their backwash.

The sorting of beach material

- The strong swash of a constructive wave moves quickly up a beach carrying sand or shingle with it.
- The coarsest material is deposited at the upper limit reached by the swash.
- The backwash carries smaller material back down the beach but it progressively loses water, and therefore energy, as it does so. This is because water passes down through spaces between the beach particles, weakening the backwash until it can only carry the lightest material.
- As the backwash continually weakens, it drops shingle and sand particles of progressively smaller size.

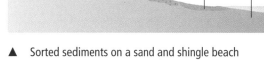

▲ Sorted sediments on a sand and shingle beach

When a storm coincides with the highest tides, large material is tossed above high tide level to the top of the beach, forming a storm ridge.

Landforms of marine deposition

Beach material eroded from cliffs further along the coast is transported along the shore by longshore drift until it is deposited in relatively calm waters of bays and inlets.

Beaches	Beaches are sand and shingle deposits between the low and high water marks (often covering a wave-cut platform). Shingle always forms a steep slope (sometimes with steps and ridges), whereas sand beaches are gently sloping. In plan, beaches can be straight, crescent shaped (in bays) and triangular or semi-circular (at the head of inlets).
Spits	These are long, narrow, low ridges of sand or shingle deposited at bends in the coast. Deposits build up and grow from a headland across a bay or river mouth, so they are attached to the land at one end and the other ends in open water. They may be re-curved at the ends because occasional strong onshore winds cause waves to approach from a different angle to that of the prevailing wind. If the spit continues to grow in the original direction each re-curve forms a hook. Some have sand dunes on them.
Bars	Like spits, except they extend right across bays or river mouths, enclosing lagoons behind them (which gradually become filled with sediment to form marshes).
Salt marshes	These are areas of low growing, salt-tolerant vegetation on mud flats. Their lower, outer parts are covered at high tide by sea water. Meandering, tidal channels (creeks) cross them and they have saltwater pools. (In the tropics and sub-tropics mangroves grow on the mud flats.) The water behind spits is sheltered and calm. Mud, brought in as the tide rises, settles there and accumulates to form mud flats on which plants tolerant both of saltwater and of being covered by water twice a day (at high tide), grow. These trap more mud and hold it firmly so the surface level rises, vegetation increases and the mud flat becomes a salt marsh or a mangrove swamp.

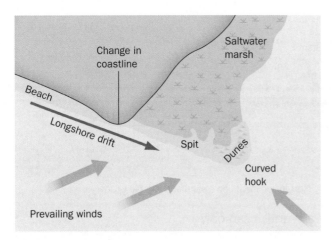

▲ How a spit forms

Practice Questions

1. Name and describe four methods of marine transportation in the diagram.

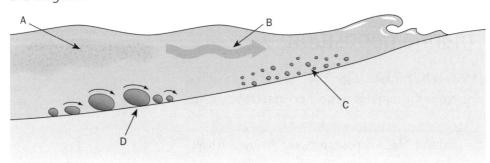

▲ The processes of marine transportation

2.

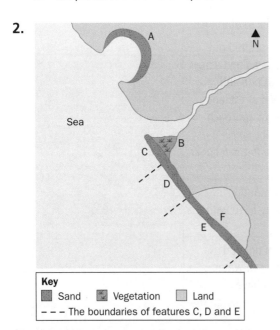

a) Identify features A to F on the diagram. (Two of them have the same identity)

b) State the direction of longshore drift along the coast.

3. Choose a section of coast that you have studied. Describe the location of different types of sediment along it.

4.

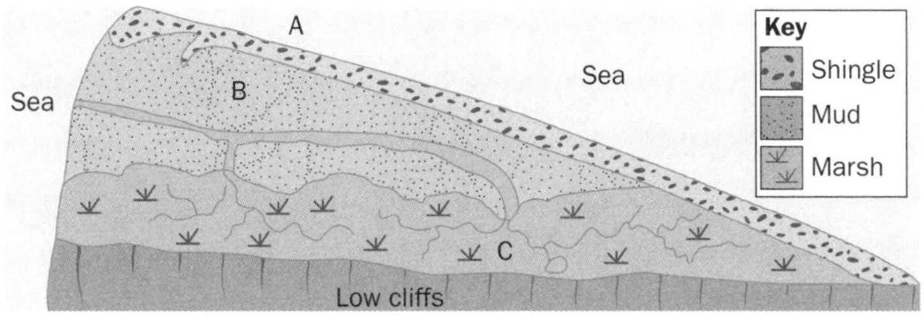

a) Name the coastal features A, B, and C shown in the diagram.
b) Describe the features of landform A (i) in plan and (ii) in cross-section.
c) Describe the features of landform C.
d) Explain the sequence of events that has changed the coastline from the time when the low cliffs were next to the sea.

5. Copy out the passage and fill in the missing words. (The size of the space does not indicate the length of the word.)

If the waves approach the shore at an _____ angle the _____ moves the sediment up the beach at an _____. The _____ brings the pebble or sand _____ down the beach. As this process is repeated the pebble or sand is moved _____ the beach. This process is known as longshore _____.

6. Match the terms with their definitions.

	Term		Definition
A	Swash	1	A long narrow ridge of sand or shingle with one end attached to the land and the other ending in open water.
B	Constructive wave	2	A ridge of sand or shingle extending from one side of a bay or estuary.
C	Bar	3	This has a stronger swash than backwash so deposits its load on the beach.
D	Groyne	4	A barrier placed across a beach to reduce sediment movement along the beach by longshore drift.
E	Longshore drift	5	A tidal channel in a salt marsh.
F	Salt marsh	6	A swampy area of water-loving vegetation on a mud flat.
G	Spit	7	The movement of sediment along a beach when the waves approach at an angle to the beach.
H	Creek	8	The forward movement of a wave up the beach.

7. Compare constructive waves with destructive waves for the following: swash, backwash, height, wave-length, frequency, result of their action.

8. Name the odd one out in each of the following lists and explain your choice:

a) Attrition, saltation, solution, suspension.
b) Marsh, mud, sand, shingle.
c) Creek, saltwater pool, shingle, vegetation.
d) High water mark, longshore current, low water mark, tidal range.

17 Coral reefs and coastal sand dunes

KEY IDEAS

→ Corals are tiny, marine animals called polyps that form reefs when millions live together in colonies.

→ Their skeletons are calcareous cups which are joined with others to form a hard, stony mass.

→ As one generation dies, the next one grows on top of it, so the reef grows upwards and outwards as the corals compete for food.

→ Coral reefs run parallel to the coast. They have breaks in them, usually at the mouths of rivers.

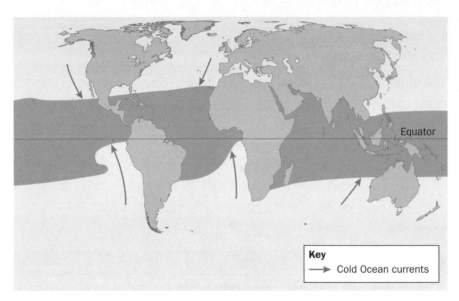

Key
⟶ Cold Ocean currents

▲ Sea with a temperature of 20° Centigrade or higher in summer and autumn. (Most coral reefs are found in these areas)

Reef-building corals cannot grow anywhere:

- They cannot live in sea temperatures lower than 18° Centigrade and grow best where the mean surface water temperature is 22–25° Centigrade. For this reason they are usually only found within 30° of the Equator and are restricted to even lower latitudes on west coasts.
- Polyps cannot survive long periods above the water, so much of the reef is at low tide level. Its outer edge is highest because that is where oxygen and food supplies are most abundant, brought by the waves that break there.
- Polyps need clean, clear, sunlit water, so cannot live where rivers deposit sediments into the sea.
- Corals grow best in conditions of high salinity.
- Few corals are found at depths greater than 30 metres because (i) they cannot live without single-celled algae which need abundant sunlight for photosynthesis and (ii) the plankton on which they feed need sunlight. Light decreases with depth. Shallow, agitated waters to about 10 metres depth are best.
- There has to be a solid surface from which the reef growth starts.

Types of coral reef

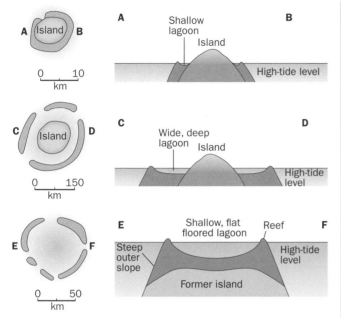

Fringing reefs e.g. on the Coral coast of Fiji	These low, narrow bands of coral are next to the coast at about low tide level. They are covered by a narrow, shallow lagoon at high tide. Beyond the lagoon, the higher outer edges of the reefs rise to about high tide level. They are parallel to the coast and their outer edges slope steeply down into the sea beyond.
Barrier reefs e.g. the Great Barrier Reef of Australia	Barrier reefs are usually several kilometres from the coast (a minimum of 500 metres) and separated from it by wide, deep lagoons below the depth at which the polyps can live. The Great Barrier Reef consists of almost 3000 reefs, separated by channels, stretching over more than 2600 kilometres.
Atolls e.g. Suvadiva Atoll in the Maldives	These are narrow, circular reefs, broken by channels. They surround a deep lagoon. Suvadiva Atoll has a lagoon more than 50 kilometres wide. The Maldives are a north to south chain of atolls, thirteen of which are very large.

Coastal sand dunes

These ridges of sand form at the back of beaches and on spits by wind deposition.

Dune ridges are moved inland because the onshore wind moves sand from their seaward sides to their lee sides. Meanwhile, new embryo dunes form nearer the sea. Eventually, lines of dunes develop parallel to the sea, with long depressions (slacks) containing water or marsh between them.

Marram grass is resistant to the drought conditions on dunes and plays a very important part in their growth because marram has a network of very long roots which help to anchor the sand.

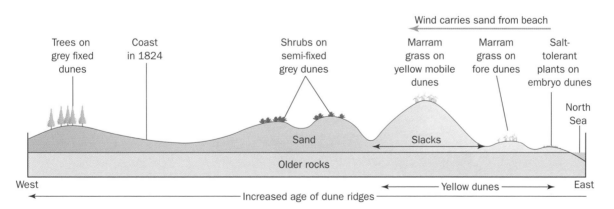

▲ The sand dune system at Gibraltar Point, England

Marram grass dies if it is trampled. The wind then removes the sand easily, sometimes cutting a valley-like shape, known as a blow-out, right through the dune.

Practice questions

1. Which **one** statement in each of the following groups is **correct**?
    a) Reef corals grow best where the sea temperatures are 22 to 25°C.
    b) Reef corals grow best where the sea temperatures are 18 to 22°C.
    c) The warmer the sea the better the coral reefs grow.
    d) Coral reefs grow best where there are cold ocean currents.

    a) The steepest part of a coral reef is nearest the coast.
    b) Coral reefs are entirely flat.
    c) Coral reefs have a steeply sloping outer edge.
    d) Coral reefs slope gently towards the sea.

2. a) Explain why most breaks in coral reefs occur opposite river mouths.
    b) Why are coral reefs described as bleached when they die and what could cause this? (Do not repeat the reason(s) given of your answer to (a)).

3. Match the terms with their definitions.

	Term		Definition
A	Atoll	1	A measure of how salty the water is.
B	Bleaching	2	Small marine animals living in calcareous cups that form coral reefs when many live together.
C	Polyps	3	A strip of coral that is parallel to the shore and separated from it by a wide, deep lagoon.
D	Reef	4	A hard calcareous stony mass of corals.
E	Salinity	5	An area of salt water near the sea but separated from the sea.
F	Lagoon	6	An area of coral reef that is close to the shore.
G	Barrier reef	7	The reef loses its colour because the algae which live with the polyps die.
H	Fringing reef	8	A circular or oval reef surrounding a deep lagoon.

4. On a copy of the diagram add labels to explain the conditions necessary for sand dunes to form.

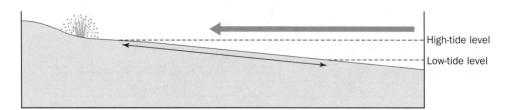

High-tide level

Low-tide level

5. Complete the crossword which contains some more challenging clues to test your knowledge about coasts.

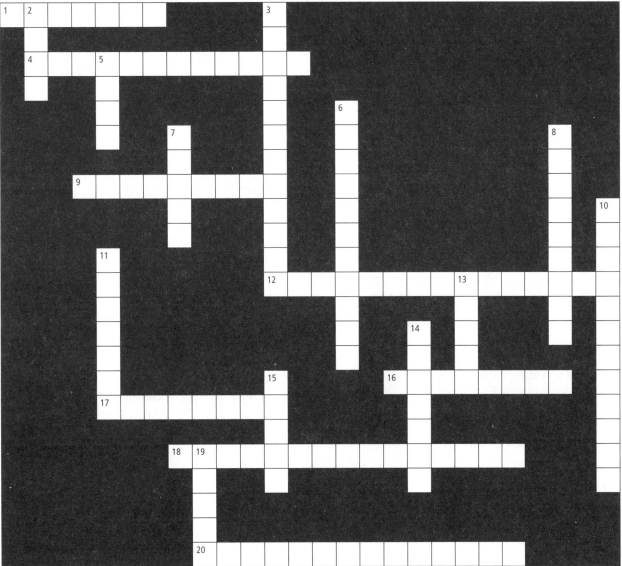

Across	Down
1 These can be found behind both coral reefs and bars. (7)	2 You would mention this between a cave and a stack in explanation. (4)
4 This type of wave is well named because it builds up beaches. (12)	3 This well named depression grows and comes before a fall. (7/5)
9 This process makes shingle round. (9)	5 The strong flow of a river can stop this coastal landform from becoming another. (4)
12 This strong wave process causes explosions of air. (9/6)	6 This feature is parallel to the coast but a considerable number of kilometres away from it. (7/4)
16 The most powerful feature of a wave on a stormy day. (8)	7 It would be very difficult and dangerous to climb this coastal feature. (5).
17 This coastal landform is composed of beach sediment but is not formed by marine processes. (4/4)	8 The bouncing movement of the sea's load. (9)
18 This well named eroded feature is what you would see if you looked down from the end of a headland at low tide. (7/8)	10 This name is given to a coral mass that is close to the coast. (8/4)
	11 Features placed on beaches to try to keep as much sand on them as possible. (7)
	13 You would need a boat to follow this through a salt marsh. (5)
	14 The force that makes the water from a wave move straight down the beach at right angles to the sea. (7)
20 This well named process moves pebbles in a zig-zag manner. (9/5)	15 This feature is usually crescent-shaped and sheltered. (5)
	19 A hard, stony, circular coastal feature that is not formed by marine erosion or deposition. (5)

Weather and weather station recording instruments

Weather is the state of the atmosphere at a particular time.

Precipitation

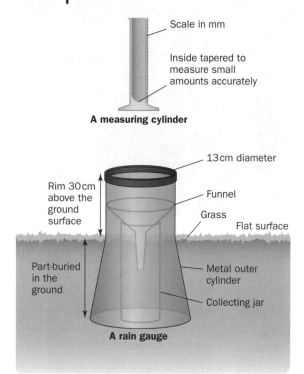

Scale in mm

Inside tapered to measure small amounts accurately

A measuring cylinder

13 cm diameter

Rim 30 cm above the ground surface

Funnel

Grass

Flat surface

Part-buried in the ground

Metal outer cylinder

Collecting jar

A rain gauge

▲   A rain gauge and how it is sited

Measuring precipitation using a rain gauge

1   At the same time each day (usually about 9 a.m.) pour the water from the collecting jar into the tapered measuring cylinder.
2   Place the measuring cylinder on a flat surface.
3   Read the water level in the measuring cylinder with the eye at the same level as the lowest part of the meniscus of the water.
4   Record the measurement. An amount too small to measure is recorded as a trace.

Temperature

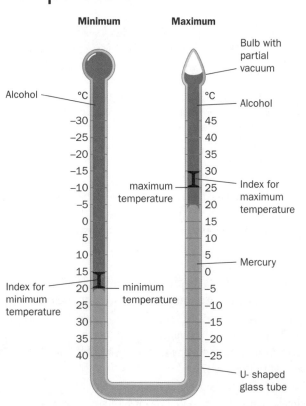

Minimum

Maximum

Bulb with partial vacuum

Alcohol

°C

−30
−25
−20
−15
−10
−5
0
5
10
15
20
25
30
35
40

°C

Alcohol

45
40
35
30
25
20
15
10
5
0
−5
−10
−15
−20
−25

maximum temperature

Index for maximum temperature

Mercury

Index for minimum temperature

minimum temperature

U-shaped glass tube

The Six's thermometer is read at eye level from the lower end of each index and reset using a magnet to draw the index back to the mercury.

▲   The Six's thermometer

Humidity

Wet and dry bulb thermometers, known together as a hygrometer, provide the readings used to find the relative humidity.

- Humidity is the amount of water vapour in a given volume of air.
- Warm air can hold more water vapour than cold air.
- Relative humidity is a measure of how much water vapour the air is holding compared with the maximum amount it could hold at its temperature.
- If the air is not saturated water evaporates from the muslin, the evaporation cools the bulb, the mercury contracts and registers a lower temperature (the depression of the wet bulb).
- Evaporation is not possible in saturated air, so both thermometers show the same temperature.

Dry bulb thermometer

Wet bulb thermometer

Wet bulb wrapped in muslin

Wick - transfers moisture to keep the muslin moist

Container of water

Air temperature

Mercury

Pressure

The metal box expands when pressure is low, and is compressed when pressure is high. In a barograph these changes are transmitted by a series of levers to a pointer which moves to the correct position on the scale on the face of the dial.

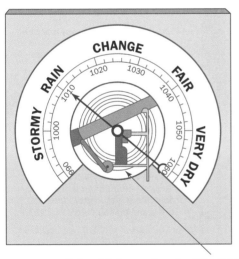

Collapsible box partly evacuated of air

▲ An aneroid barometer

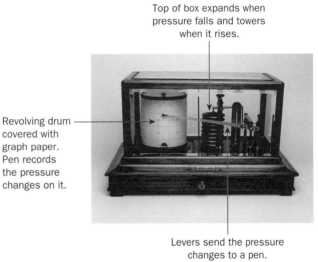

Top of box expands when pressure falls and towers when it rises.

Revolving drum covered with graph paper. Pen records the pressure changes on it.

Levers send the pressure changes to a pen.

▲ A barograph

Wind

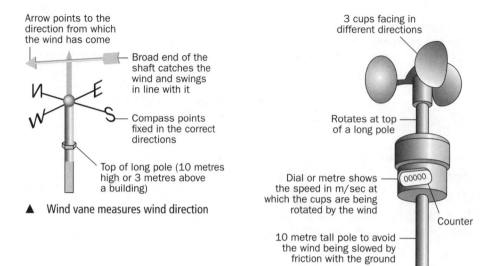

Arrow points to the direction from which the wind has come

Broad end of the shaft catches the wind and swings in line with it

Compass points fixed in the correct directions

Top of long pole (10 metres high or 3 metres above a building)

▲ Wind vane measures wind direction

3 cups facing in different directions

Rotates at top of a long pole

Dial or metre shows the speed in m/sec at which the cups are being rotated by the wind

00000

Counter

10 metre tall pole to avoid the wind being slowed by friction with the ground

▲ Anemometer measures wind speed

Meteorological stations use automated digital recording instruments that transmit data to computer screens. Digital handheld instruments show the readings on a screen.

The Stevenson screen

• All the thermometers are housed in this wooden box, often with the barometer. It is designed to ensure that the instruments give the correct readings.

Weather station layout

• Thermometers must be on grass and away from buildings which may radiate heat.
• Wind vanes and anemometers should be in an open space away from any trees or buildings, at least three times the height of the nearest obstacle away from it.
• The rain gauge must be in an open space with the distance from the nearest object twice its height.
• Barometers and barographs should be kept away from strong air movements, direct sunlight and heat sources.

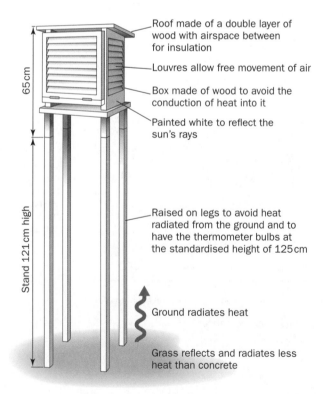

Roof made of a double layer of wood with airspace between for insulation

Louvres allow free movement of air

Box made of wood to avoid the conduction of heat into it

Painted white to reflect the sun's rays

65 cm

Stand 121 cm high

Raised on legs to avoid heat radiated from the ground and to have the thermometer bulbs at the standardised height of 125 cm

Ground radiates heat

Grass reflects and radiates less heat than concrete

Practice questions

1. Fill in the blanks to explain how a Six's thermometer works.

   When the temperature rises some _____ in the right arm evaporates into the space in the _____ and the alcohol in the left arm _____ and pushes the _____ up the right arm. This pushes a metal _____ which is left at the _____ temperature reached.

   When the temperature falls, the _____ cools and _____. The mercury moves up the left arm, pushing the _____ index with its _____ .

2. Name weather recording instruments with each of the following features:
   a) cups
   b) funnel
   c) levers
   d) arrow
   e) corrugated metal box
   f) wick

3. Match the beginnings with their correct endings. The site of a rain gauge must be…

	Beginnings		Endings
A	Away from buildings…	1	to avoid drips of rain making the amount caught higher than it should be.
B	Away from trees…	2	for stability and to prevent evaporation of the rain in the collecting jar.
C	On grass, with its rim 30 cm above the ground…	3	to keep the correct diameter of the gauge.
D	Part buried…	4	to prevent splash adding to the rain in the gauge. (In areas with hard surfaces and no grass the rain gauge has to be raised on a tripod.)
E	Standing vertically upright…	5	to avoid shelter reducing the rain collected.

4. State the advantages of using digital instruments.

5. Fill in the spaces to complete the account of the formation of rainfall.

   When air temperature increases it _____ water from water surfaces such as seas, lakes, rivers and vegetation, so increasing its _____. When the air _____ the amount of water vapour it can hold decreases. If the air is cooled sufficiently, it will reach the temperature at which it is holding the maximum amount of water vapour that can be held at that _____. That is called the _____ point and the air is said to be saturated (in a state of saturation), with a _____ humidity of 100%. Any further cooling results in _____ . The excess water vapour changes into _____ droplets or ice, depending on the temperature of the air.

6. Match the terms with their definitions.

	Term		Definition
A	Louvre	1	A mixture of gases that encircle the Earth, about 99% nitrogen and oxygen with small amounts of other gases such as water vapour, methane and ozone.
B	Trace	2	The change in state from a liquid, such as water, to a gas, such as water vapour, because of heating.
C	Saturation	3	The state of the air when it is holding the maximum amount of water vapour for its temperature.
D	Millibar	4	The amount of water vapour in a given volume of air at a particular time.
E	Insulation	5	Protection from heating and cooling by outside elements.
F	Evaporation	6	A narrow space to allow air to enter and leave freely.
G	Meniscus	7	The condition of the atmosphere at any particular time.
H	Atmosphere	8	An amount of precipitation too small to measure.
I	Weather	9	The unit used for pressure measurement.
J	Humidity	10	The curved outer surface of a liquid which is convex in mercury and concave in water. A reading for mercury is taken at the highest part of it and for water at the lowest part of it.
K	Water vapour	11	The amount of water vapour the air is holding expressed as a percentage of the maximum amount it could hold at a certain temperature.
L	Dew point	12	Liquid and solid ice particles that fall from the atmosphere to the Earth's surface. (Rain and drizzle are in liquid form while snow and hail are solid ice. Sleet is partially melted ice.)
M	Precipitation	13	The invisible gaseous form of water which is present in all air in varying quantities.
N	Relative humidity	14	The temperature at which air is holding the maximum amount of water vapour for its temperature.

7. Give a reason for each of the following design features of the Stevenson screen:

a) The door faces north in the northern hemisphere and south in the southern hemisphere.

b) Louvres allow air into and out of the screen.

c) The screen is painted white.

8. Give the term for the following descriptions:

a) An instrument which measures the pressure of the air and makes a continuous recording of it on graph paper.

b) The process by which a gaseous substance, such as water vapour, changes into a liquid, such as water, because of cooling.

c) The weight of the atmosphere at any place.

d) The degree of heat or coldness of the air, sea or other body.

e) The components of weather, including temperature, wind speed and direction, humidity, precipitation and pressure.

f) A place where instruments are used to measure all aspects of the weather.

Weather data calculations, graphs and diagrams

Calculations

Rainfall:

- monthly and annual rainfall totals,
- mean (average) monthly and annual rainfall, over a minimum of thirty years.

Describing annual rainfall amounts

Annual rainfall in millimetres	Description of the amount
0 – 249	Very low
250 – 499	Low
500 – 999	Moderate
1000 – 1999	High
Over 2000	Very high

Temperature:

- daily range of temperature
- mean daily temperature
- mean monthly temperature
- annual range
- mean annual temperature

Describing temperatures	
Temperature (°C)	Description
30° and above	Very hot
20 – 29	Hot
10 – 19°	Warm
0 – 9	Cool
−10 to −1	Cold
Below −10	Very cold

Describing temperature ranges	
Temperature range	Description
0 – 3°C	Very small
4 – 8°C	Small
9 – 19°C	Moderate
20 °C and above	Large

Pressure

Pressure is not totalled over time. Mean sea level pressure is 1013 millibars. However, a higher pressure area surrounded by lower pressures would be described as a high pressure system, even if it did not reach 1013 mb. Low pressure areas surrounded by higher pressures can have central pressures of more than 1013 mb.

Wind

It is important to know where the wind has come from because it brings characteristics of the temperature and moisture of the areas it has passed over to influence the weather where it is recorded. The most frequently occurring wind in an area is known as the prevailing wind and the wind direction giving the strongest winds is known as the dominant wind.

Always express wind direction clearly: 'it is a westerly wind' or 'the wind is from the west' cannot be misinterpreted (whereas 'in a westerly direction' is ambiguous).

Describing wind speeds	
Wind speed (km/hour)	Description of the wind speed
Below 50	Calm, light, moderate or strong winds
50 – 100	Gale
101 – 118	Storm
119 and above	Hurricane

Relative humidity

Knowing how near the air is to being saturated is vital for accurate forecasting of precipitation. It is found by looking in a relative humidity table where the depression of the wet bulb thermometer line intersects with the dry bulb temperature line.

Dry bulb (°C)	Depression of wet bulb (°C)					
	1	2	3	4	5	6
14	90	79	69	60	51	41
16	90	81	71	62	54	45
18	91	82	73	65	57	49
20	91	82	74	66	58	51
22	92	83	75	68	60	53
24	92	84	77	70	63	56
26	92	85	77	71	64	57

◄ Part of a relative humidity table

Graphs and diagrams drawn from weather recordings

Climate graphs

These show mean monthly temperatures and mean monthly precipitation calculated over at least 30 years.

Keep the following points in mind when describing climate graphs:

- Equatorial climate stations have no seasons but it is important to refer to summer and winter when describing climate graphs of other locations.
- Summer is the period with the hottest months and winter the period with the coldest months. (Spring and autumn are not usually mentioned).
- Do not give a month by month account of the changes in temperature but quote figures and describe in words the highest and lowest temperatures, mean and range.

Remember to describe both the total and seasonal distribution of the rainfall.

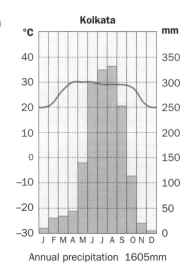

Kolkata

Annual precipitation 1605mm

Dispersion diagrams are useful for showing distributions. This diagram shows the distributions of annual rainfall totals for the last 13 years in two places.

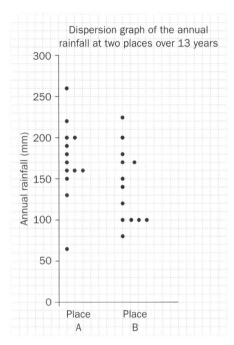

Dispersion graph of the annual rainfall at two places over 13 years

The range, median (middle number) and modal (most frequently occurring) values can be quickly read and compared. The position of the calculated mean can be shown on the diagram.

Wind roses

Bars radiate from a central area for each compass direction. The length of each bar shows the number of days with the wind from that direction over a period of time (usually a month). The number of calm days is written in the centre.

Isoline and choropleth maps

- isohyets are lines joining places having the same rainfall,
- isotherms join places with the same temperature,
- isobars join places with the same pressure,
- isoline maps become choropleth maps when shaded between the isolines. Conventional shading is heaviest for the largest value and progressively lighter with decreasing values.

Weather maps (Synoptic charts)

Readings taken at meteorological stations are plotted on synoptic charts.

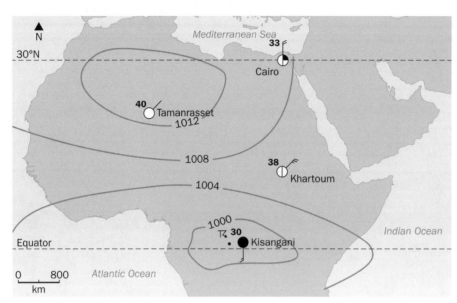

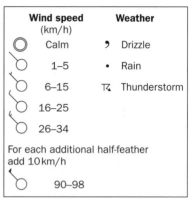

▲ A synoptic chart for part of Africa at 2 p.m. on a day in March

Practice questions

1. State the readings shown on the instruments in Chapter 18:

   a) From the Six's thermometer (Page 72) give the actual temperature, and the maximum and minimum temperatures. Calculate the range and mean daily temperatures.

   b) State the wind direction shown by the wind vane on page 74.

   c) Use the table on page 78 to state the relative humidity when the dry bulb thermometer reading is 22°C and the wet bulb thermometer reading is 18°C.

   d) What does a small wet bulb depression indicate?

2.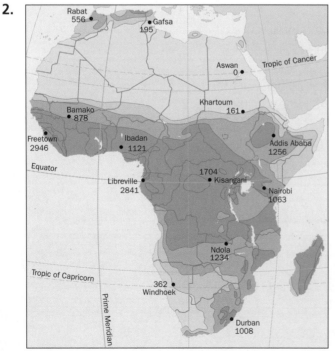

   a) State and describe the annual precipitation at (i) Aswan and (ii) Kisangani,

   b) Describe the general distribution of annual precipitation over Africa of areas having (i) under 500 mm and (ii) areas having more than 1000 mm.

   c) What type of map is this?

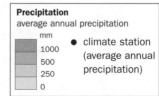

3. Look at the climate graph for Kolkata (page 78).

   a) In which hemisphere is Kolkata and how does the graph show this?
   b) Describe its mean annual temperature, annual range of temperature and seasonal temperatures.
   c) Describe the annual total and seasonal distribution of the rainfall.

4. a) Look at the dispersion diagram on page 79. For both Place A and Place B list:

   i) the difference in values between the wettest and driest years,
   ii) the modal rainfall,
   iii) the median value.

   b) Which place has the most unreliable rainfall?

5. Match the terms with the definitions:

	Term		Definition
A	Isohyet	1	Plots at the correct levels against a vertical scale.
B	Dispersion diagram	2	A map of recordings at weather stations over an area.
C	Isotherm	3	Line joining places with the same temperatures.
D	Prevailing wind	4	Line joining places with the same pressure (reduced to sea level).
E	Isobar	5	The wind direction from which the wind comes most frequently.
F	Dominant wind	6	Line joining places with the same amount of rainfall.
G	Synoptic chart	7	The direction from which the strongest winds arrive.

6. State the formula for calculating each of the following:

   a) the daily temperature range,
   b) the mean daily temperature,
   c) the mean monthly temperature,
   d) the annual temperature range,
   e) mean annual temperature.

7. How are the following shown on a synoptic chart?

   a) Pressure
   b) A weather station
   c) The extent of cloud cover
   d) The wind direction
   e) Wind speed
   f) Precipitation
   g) Temperature

8. a) Why is temperature on a climate graph shown as a line graph whereas rainfall is in bars?

   b) Why are pressures converted to their sea level equivalents when plotted on maps?

20 Clouds and weather hazards

Cloud types and extent

Clouds consist of tiny water droplets or ice particles which are too light to fall to Earth. They form when air rises and cools until water vapour condenses into water droplets or, if it is sufficiently cold, to ice crystals.

Air continues to rise while it is warmer and lighter than the air into which it is rising. It cannot rise above the tropopause. Clouds with the greatest vertical extents form in the tropical zone where the tropopause is at its highest.

Clouds only produce precipitation if they have a lot of water or ice particles so that the particles can collide and join together. Precipitation occurs if the particles grow heavy enough to fall through the rising air currents. The only clouds that produce precipitation are nimbostratus which are thick enough, cumulonimbus which have a great vertical extent and strongly rising air currents, and stratus from which drizzle may fall.

The main types of cloud

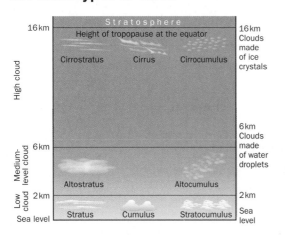
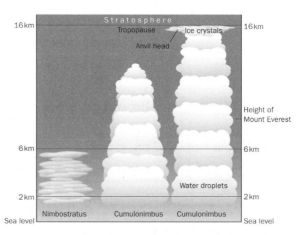

Cloud	Level	Description	Associated weather
Cirrus	High (above 6 km)	This cloud is thin, white and made of ice crystals. It forms narrow wisps or threads (cirrus means hair-like). It can also be feather-like in shape.	Fine
Cirrostratus	High (above 6 km)	This cloud is a thin, white layer made of ice crystals, with a wide horizontal extent. It often covers the whole sky.	Fine
Cirrocumulus	High (above 6 km)	This cloud is thin, white, made of ice crystals and slightly heaped.	Fine
Altostratus	Medium (2-6 km)	This cloud is a layer of water droplets, which can be thin and white or thick and grey	Fine
Altocumulus	Medium (2-6 km)	This is a heaped cloud of water droplets, whch can be white or thick enough to look light grey.	Fine
Stratus	Low (0-2 km)	This cloud is a thin, uniform, grey sheet of small water droplets, with a fairly flat base.	It may be thick enough to produce drizzle
Cumulus	Low (0 -2 km)	This cloud is white with a darker, flat base and globular upper surface. It is made of water droplets. It may have a small or considerable vertical extent.	Sunny by day, fine weather
Stratocumulus	Low (0-2 km)	This is a layer of cloud with some heaped sections, giving white and grey parts. It is made of water droplets.	Fine
Nimbostratus	The base can be low, or above 2 km	This cloud is a thick, dark grey layer of water droplets.	Steady rain or drizzle
Cumulonimbus	A low base, but the cloud extends up to high levels	A dense, dark grey cloud with a great vertical extent. It grows from a cumulus cloud to have a high, billowy head (or a flat top if it reaches the tropopause). If it then spreads out, it has an anvil top. It is composed of ice crystals at the top and water droplets at lower levels.	Very heavy rain, or snow showers, often with hail and thunder and lightning

How cloud extent is measured

The extent of cloud cover is estimated by eye and expressed in the number of oktas (eighths) of the sky covered with cloud.

Weather hazards

Tropical storms

The most intense tropical storms have wind speeds of at least 119 kilometres an hour. These deep low pressure systems with ferocious spirals of air are known as

- hurricanes in the Caribbean Sea, Gulf of Mexico and west coast of Mexico;
- cyclones in the Indian Ocean, Bay of Bengal and northern Australia;
- typhoons in the South China Sea and west Pacific Ocean.

These are shown on the map on page 100 of chapter 24.
They have destructive winds and intense rainfall on either side of the central eye of the storm and cause storm surges – rapid rises of sea level.

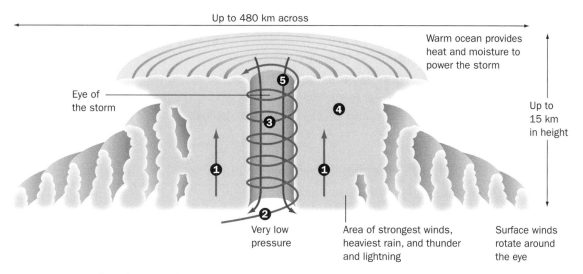

▲ Cross-section through a tropical storm

The weather in the eye of the storm is calm and sunny.

Drought

Drought is a longer than usual period of dry weather. Droughts occur when rain fails at a time when it is normally expected to fall and so they cause problems for vegetation and human activities. The problems become acute when the rains fail in successive years. Droughts can occur almost everywhere but there are areas of the world where droughts are particularly severe and frequent, such as in the Sahel of Africa (Map p.103.).

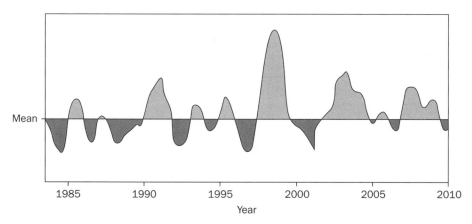

▲ Annual rainfall variations in central Kenya above and below the mean value

Practice questions

1. Which of the following is the odd one out? Give your reason for each.
    a) Cirrus, cumulus, cirrocumulus, cirrostratus.
    b) Cumulus, cumulonimbus, cirrocumulus, stratocumulus.
    c) Nimbostratus, cumulonimbus and altocumulus.

2. a) Use the map on page 100 to describe where tropical storms form.
    b) When is there a brief spell of sunshine during the passage of a tropical storm over a place?

3. Which **one** statement in each of the following groups is correct?
    a) All types of cloud are made of water droplets.
    b) Clouds only form when air is rising.
    c) Most types of cloud produce rain.
    d) Nimbostratus is the tallest cloud.

    a) Tropical storms are known as hurricanes in the Bay of Bengal.
    b) Tropical storms are known as cyclones in the South China Sea.
    c) Tropical storms have wind speeds of at least 119 km an hour.
    d) There is one spell of very intense rain and strong wind as a tropical storm passes.

    a) Africa is the continent with the biggest total area prone to droughts.
    b) Hurricanes form very near the Equator.
    c) Tropical storms mainly affect the west coasts of continents.
    d) A drought is the same as a dry season.

    a) Cirrocumulus is a middle layer cloud.
    b) Tropical storms form in summer and autumn.
    c) Typhoons have a period of rain as the eye of the storm passes.
    d) Wind direction remains the same as a tropical storm passes.

4. Why do some clouds appear to be black while others are white?

5. What do the following ways of describing clouds mean?
    a) Stratus
    b) Cumulus
    c) Cirrus
    d) The prefix cirro
    e) The prefix alto

6. Use a number and units to describe:
    a) full cloud cover
    b) no cloud cover
    c) half the sky covered with cloud.

7. Complete the crossword.

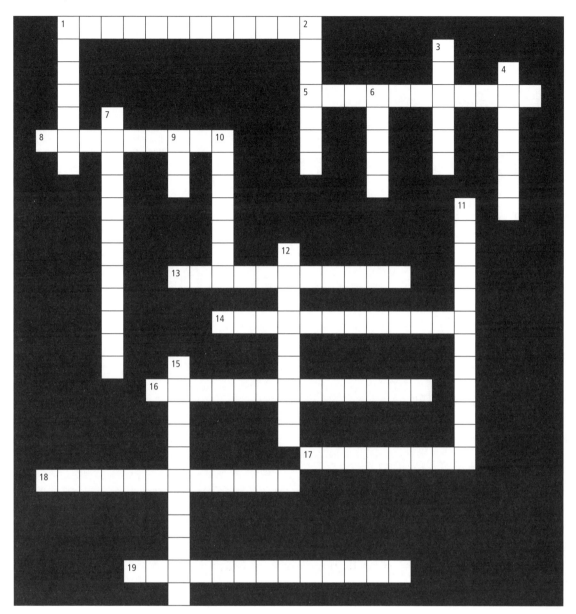

Across	Down
1 This is the cloud from which the most precipitation falls. (12)	1 This is the name used for an intense low pressure system in the Indian Ocean and Bay of Bengal. (7)
5 Layer cloud formed at a medium level. (11)	2 Low level layer cloud, from which drizzle might fall. (7)
8 The description of the shape at the top of a cloud with strong rising air currents that reaches the tropopause. (5/4)	3 The wispiest cloud. (6)
	4 A low level cloud with a flat base and globular upper surface. (7)
13 This cloud is medium level with a flat base and globular top. (11)	6 Clouds are measured in these. (5)
14 A small, high level, globular cloud. (12)	7 This stratus cloud has sufficient vertical height to produce rain. (12)
16 This name describes any low pressure area in low latitudes with wind speeds of 63 km per hour and above. (8/5)	9 The central part of a tropical storm which gives a short calm, sunny and dry period. (3)
	10 A long period without rain, or with little rain, when more rain would normally be expected to fall. (7)
17 These bring devastation to the South China Sea and Philippines. (8)	11 This high level cloud is a very thin layer. (12)
18 This is essential for clouds to form. (12)	12 An intense tropical storm in the Caribbean and Gulf of Mexico. (9)
19 A layer or line of mainly attached cumulus cloud. (13)	15 The zone of weather, the part of the atmosphere nearest the Earth's surface. (11)

Climate of tropical rainforest areas

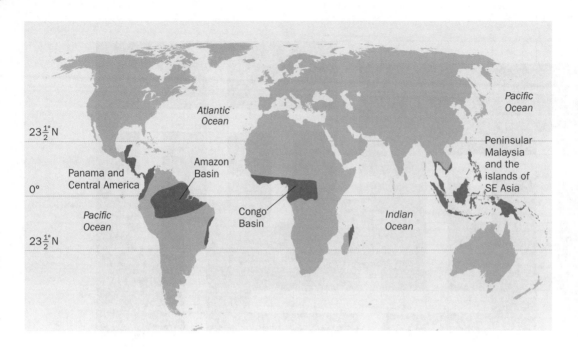

Tropical rainforests grow in three large areas with an Equatorial climate, all within 10° of the Equator and all lowlands.

The equatorial climate

The climate is remarkably uniform with no seasons. It is hot and wet all year. Its characteristics are:

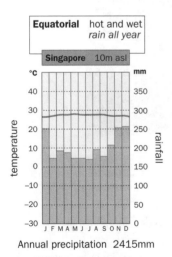

Annual precipitation 2415mm

▲ Climate of Singapore

- hot with a mean annual temperature of about 27°C.
- a low annual range of temperature of 2 or 3°C.
- a higher diurnal (daily) range of about 7°C.
- high annual rainfall (over 1500 mm).
- at least 60 mm of rainfall in the drier months.
- high relative humidity.
- light and variable winds.
- a lot of cloud.

Influences on the temperatures

Latitude

The orbit of the Earth round the sun causes varying angles of the midday sun at different times in the year because it is tilted on its axis.

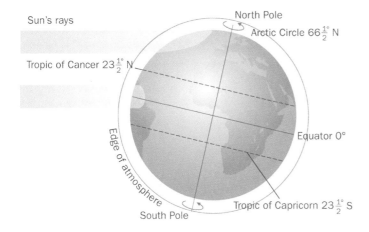

The lower the latitude, the higher the angle of the sun's rays and the greater the heating because:

- high angle rays fall on a smaller area, so the heat is more concentrated than at higher latitudes.
- the rays have a shorter passage through the atmosphere so less insolation (incoming solar radiation) is reflected and scattered back to space and more reaches the Earth's surface.

Cloud

Places near the Equator are not the hottest places on Earth because, during the day, clouds reduce surface temperatures by reflecting some of the solar radiation back to space from their white upper surfaces and by absorbing some of it. About 50% gets through to be absorbed by the Earth's surface, converted into long-wave radiation and radiated back to space at night. Clouds absorb this and re-radiate it back to Earth. So the cloudy nights of the Equatorial climate keep it warmer than it would otherwise be.

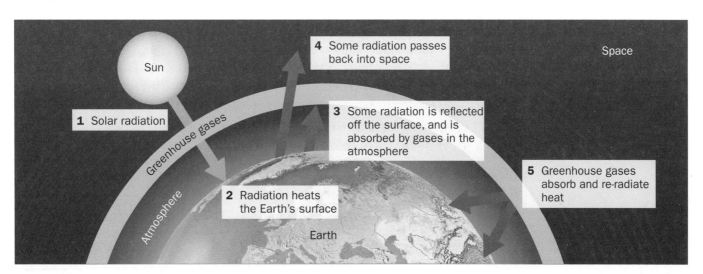

Water droplets in clouds, water vapour, carbon dioxide and other greenhouse gases keep the air near the Earth's surface warmer by allowing more incoming solar radiation to pass through the atmosphere than outgoing radiation from the Earth.

Reasons for the high rainfall

The combination of high temperatures and of air with a high moisture content results in convectional rainfall.

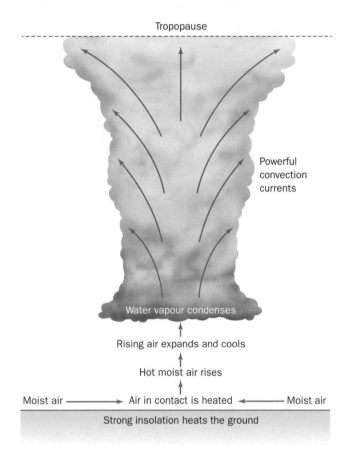

Tropopause

Powerful convection currents

Water vapour condenses

↑
Rising air expands and cools

↑
Hot moist air rises

↑
Moist air ——→ Air in contact is heated ←—— Moist air

Strong insolation heats the ground

These towering cumulonimbus clouds produce very heavy rain, often with thunder and lightening.

As air is rising, surface winds are light and low pressure results. Also, warm air expands, giving less weight in a column of warm air than in colder air. The equatorial zone is a permanent low pressure belt.

- As winds blow into low pressure to replace the rising air, they meet and rise in the Inter-Tropical Convergence Zone (ITCZ).
- The ITCZ moves into the hemisphere where the sun is overhead.

Practice questions

1.

Month	Jan	Feb	Mar	Apr	May	Jun	Jul	Aug	Sept	Oct	Nov	Dec
Mean temp. (°C)	26.5	27	27.5	27.5	28	28	28	28	28	27.5	27	26.5
Mean max. temp. (°C)	30	31	31	31	32	32	32	32	32	31	31	30
Mean min. temp. (°C)	23	23	24	24	24	24	24	24	24	23	23	23
Mean relative humidity (%)	82	79	79	81	81	79	80	80	80	80	82	82
Mean rainfall (mm)	252	169	190	183	175	175	170	197	179	214	253	258
Mean daily sunshine hours	5.1	6.4	6.1	5.9	5.9	6.2	6.2	6.0	5.6	5.3	4.6	4.5

▲ Climate statistics for Singapore (1° 23′ N, near sea level)

Use the table of Singapore's climate statistics to answer the following:

a) Estimate and describe the mean annual temperature.

b) State the mean annual temperature range. (Show your working.)

c) Is this a cloudy climate? Explain your answer.

d) Estimate and describe the mean annual relative humidity.

e) Describe the total annual precipitation of 2415 mm and how it is distributed through the year.

2.

Jan	Feb	Mar	Apr	May	Jun	Jul	Aug	Sept	Oct	Nov	Dec
60	80	175	159	132	104	136	169	186	121	195	83

▲ The mean monthly rainfall at Kisangani in millimetres

a) Describe how the rainfall at Kisangani is typical of an Equatorial climate.

b) Explain why the air in Equatorial climates is very moist.

3. Select the correct words from the options to complete each statement.

Statement	Options
The sun is overhead at the Tropic of Capricorn on ….	March 21st, June 21st, September 23rd, December 22nd.
Pressures in Equatorial latitudes are ….	High, low, moderate.
In Equatorial climates the daily temperature range is ……….. than the annual temperature range.	Larger, smaller, the same as.
Aspect means …	The height of a place, the way a slope faces, nearness to the sea.
Climate is the average of the weather over at least …years.	10, 20, 30, 40, 50,100.
Convectional rainfall results from air being forced to rise …	Because of contact with a hot land surface, over mountains, by cooler air beneath pushing it up.

4. Fill in the spaces to explain why places with an the Equatorial climate are always hot. The missing words are all names for lines of latitude or months.

The sun at noon is directly overhead at the _____ on 21_____ and 23 _____ and very nearly overhead during the rest of the year.

The sun is furthest from overhead on 21_____, when it is overhead at noon at the _____(23½°N) and on 22 _____ when it is overhead at the _____, (23½°S). On these dates it is still at a high angle (66½°).

5. Explain the following characteristics of areas with an Equatorial climate:

   a) Convectional rain usually falls in the afternoon
   b) Places near the Equator have a double rainfall maxima (two periods of higher rainfall) in a year.

6. The following statements are about places with an Equatorial climate. Which **one** statement is **correct**?

   a) Mean annual temperatures are about 30°C.
   b) The minimum monthly rainfall is 60mm.
   c) Relative humidity is low.
   d) Winds are strong.

Climate of tropical desert areas

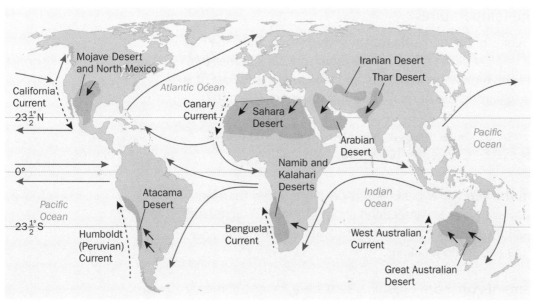

Key
- ----▶ Cold ocean current
- ⟶ Warm ocean current
- ⟶ Trade winds
- ▨ Tropical deserts

Notes:
The Kalahari Desert
is a semi-desert area to the
east of the coastal Namib Desert.

The winds are the arrows over
the land masses.

Most tropical deserts are located astride the tropics, between latitudes 15° and 30°, on the western sides of continents. However, some, such as the Mojave and Thar deserts, lie between the Tropic of Cancer and 40°N.

Month	Jan	Feb	Mar	Apr	May	Jun	Jul	Aug	Sep	Oct	Nev	Dec	Year
Mean temp. (°C)	30.8	33.0	36.8	40.1	41.9	41.3	38.4	37.3	39.1	39.3	35.2	31.8	37.1
Mean max. temp. (°C)	23.2	25.0	28.7	31.9	34.5	34.3	32.1	31.5	32.5	32.4	28.1	24.5	29.9
Mean min. temp. (°C)	15.6	17.0	20.5	23.6	27.1	27.3	25.9	25.3	26.0	25.5	21.0	17.1	22.7
Rainfall (mm)	0	0	0	0	4	5	46	75	25	5	1	0	161

▲ Climate of Khartoum (380 metres above sea level) in the Sahara Desert

Tropical desert climates are hot and dry all year. Their characteristics are:

- Very hot summers
- Hot or warm winters
- Moderate annual ranges
- Very high daily temperature ranges
- Very low rainfall - below 250 mm
- Occasional torrential rain – rain is rare and erratic (some places have no rain for years)
- Strong, constant, dry Trade winds (NE Trades in the northern hemisphere and SE Trades in the southern)
- High sunshine totals
- Low relative humidity
- High pressure

Influences on the temperatures

1. **Latitude** The tilting of the hemispheres towards or away from the sun causes the tropical desert climate to have seasons. In summer the noonday sun is at a high angle in the sky and in winter it is lowest. It is never very low so winters are normally hot and summers very hot.

2. **Altitude** Air temperature decreases as altitude increases. The rate of decrease averages 0.6 °C for every 100 metres.

3. **Distance from the sea** Water heats and cools more slowly than land so coastal areas have warmer winters and cooler summers than inland places. This maritime influence results in annual temperature ranges being smaller than at inland locations.

4. **Cold ocean currents offshore** Coastal deserts have lower summer temperatures than expected for their latitudes. The cold ocean currents off the west coasts of tropical deserts lower the temperatures of the coasts because onshore winds are chilled by contact with them.

5. **Lack of cloud** Desert air has very low relative humidity, so little cloud forms. This causes extreme diurnal (daily) temperatures and very large daily temperature ranges all year round.

6. **Aspect** North of the Tropic of Cancer, north-facing slopes are cooler than south-facing slopes because they face away from the sun. In the Southern Hemisphere south of the Tropic of Capricorn, the north-facing slopes are the warmer.

Influences on the precipitation

1. High Pressure

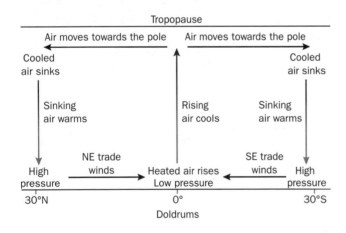

The Trade winds are strong, constant and dry. Sinking air and dry offshore Trades cause the very low precipitation of the deserts. The directions of the Trade winds result from

*   winds blowing out of the high pressure areas into the equatorial low pressure zone
*   winds being deflected by the Earth's rotation, to the right in the Northern Hemisphere and to the left in the Southern Hemisphere.

2. **Cold ocean currents offshore** Coastal deserts have very little rain because cooling causes condensation over the cold current, forming fog. This removes moisture from the air moving inland.

3. **Relief** Relief rainfall is caused by winds crossing high ground.

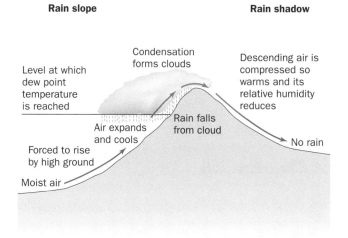

▲ Relief rain and rain shadow

4. **Temperature** Occasional convectional storms occur in the summer heat, especially on the desert margins nearest the Equator.

Practice questions

1. Look at the graph showing the climate of Tamanrasset on page 106, chapter 25.

   a) In which hemisphere is Tamanrasset and how does the climate graph indicate this?

   b) Describe the total annual rainfall and its distribution.

   c) Calculate by how much Tamanrasset's temperature is reduced by its altitude of 1377m above sea level.

2.

Month	Jan	Feb	Mar	Apr	May	Jun	Jul	Aug	Sep	Oct	Nov	Dec
Mean temp. (°C)	19	18.5	18	17	16	15.5	14.5	14	15	16	17	18
Mean max. temp. (°C)	22	22	22	21	20	20	19	18	19	20	21	22
Mean min. temp. (°C)	16	15	14	13	12	11	10	10	11	12	13	14

   a) How do the mean monthly temperatures of this desert weather station differ from those of Khartoum? (See page 91)

   b) Suggest 3 possible reasons for the differences.

3. Calculate the mean daily temperature range for April in Khartoum and describe it.

4. Rearrange the letters in the anagrams to find three influences on climate.
   a) onisaltino
   b) trine-practilo vengenoccer enzo
   c) eflier flarnial

5. Explain the following characteristics of areas with a Tropical Desert Climate:
   a) Mountains and plateaux in deserts are cooler than the surrounding lowlands,
   b) Daytime temperatures are often as high as 38°C in the shade and can reach 50°C in summer,
   c) At night temperatures drop rapidly to about 15°C in summer and 5°C in winter.

6. Fill in the blank spaces to explain why the low precipitation of tropical deserts is linked to pressure. Use the diagram of the Hadley Cell (page 92).

   The rising air that causes _____ in Equatorial climates eventually _____ to the Earth's surface at about 30°N and 30°S, resulting in _____ pressure at the surface. Sinking air warms, its relative humidity _____ so deserts are arid. This dry air moves back to Equatorial latitudes as the _____ winds.

The tropical rainforest and tropical desert ecosystems

An ecosystem is an area in which plants and animals live in balance with their environment and are inter-linked with it.

The relationship between the tropical rainforest vegetation and the soil

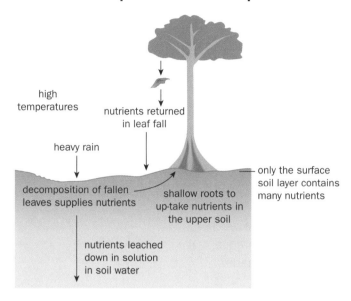

- high temperatures
- heavy rain
- nutrients returned in leaf fall
- only the surface soil layer contains many nutrients
- decomposition of fallen leaves supplies nutrients
- shallow roots to up-take nutrients in the upper soil
- nutrients leached down in solution in soil water

The soils are red because there is a lot of iron near the surface.

Tropical rainforest

Tropical rainforests need a mean annual temperature of 24° C and a minimum annual rainfall of 1500 mm.

Rainforests are so dense and continuous that light does not penetrate far into them except by rivers and clearings. The forest is well adapted to the climate. It has a five tier structure.

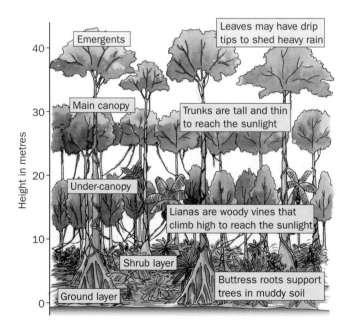

Height in metres

- Emergents
- Leaves may have drip tips to shed heavy rain
- Main canopy
- Trunks are tall and thin to reach the sunlight
- Under-canopy
- Lianas are woody vines that climb high to reach the sunlight
- Shrub layer
- Buttress roots support trees in muddy soil
- Ground layer

Other characteristics:

- There are a great number of tree species in an area but they all look alike.
- Each species is widely spread apart.
- They are mainly hardwoods, such as ironwood and mahogany.

The relationship of the natural vegetation and the climate

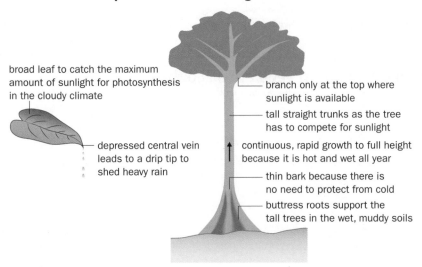

broad leaf to catch the maximum amount of sunlight for photosynthesis in the cloudy climate

depressed central vein leads to a drip tip to shed heavy rain

branch only at the top where sunlight is available

tall straight trunks as the tree has to compete for sunlight

continuous, rapid growth to full height because it is hot and wet all year

thin bark because there is no need to protect from cold

buttress roots support the tall trees in the wet, muddy soils

The forest is not seasonal; some trees will have flowers and others fruit while some are losing their leaves because the climate has no seasons. So, even though the trees are deciduous, the forest has an evergreen appearance. A tree may have branches with no leaves while others have full foliage.

The value of tropical rainforests in maintaining a healthy ecosystem

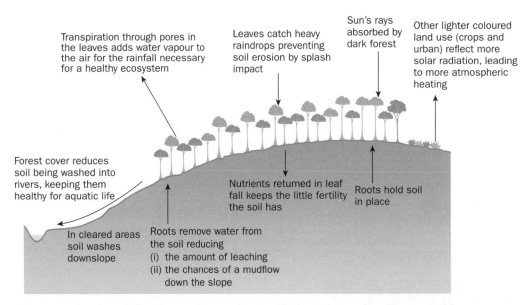

Transpiration through pores in the leaves adds water vapour to the air for the rainfall necessary for a healthy ecosystem

Leaves catch heavy raindrops preventing soil erosion by splash impact

Sun's rays absorbed by dark forest

Other lighter coloured land use (crops and urban) reflect more solar radiation, leading to more atmospheric heating

Forest cover reduces soil being washed into rivers, keeping them healthy for aquatic life

In cleared areas soil washes downslope

Roots remove water from the soil reducing
(i) the amount of leaching
(ii) the chances of a mudflow down the slope

Nutrients returned in leaf fall keeps the little fertility the soil has

Roots hold soil in place

Forests maintain soil fertility, water quality and help to keep the climate stable. During photosynthesis plants take in carbon dioxide, a greenhouse gas. They also respire, releasing oxygen, a vital gas for human life. Forests, acting as carbon sinks, lessen the effects of enhanced global warming.

Animals in the tropical rainforest ecosystem

The tropical rainforest has a very rich and diverse animal life, because it provides a variety of **habitats** and an abundance of vegetation for food. The forests of Borneo contain: over 200 mammal species, over 400 bird species, 100 amphibian species, nearly 400 fish species, and literally thousand of insect species. New animal species are discovered every year in the rainforests of Borneo and elsewhere.

Each layer of the forest has different conditions of sunlight, temperature and moisture. Examples of animals living on the forest floor are pygmy elephants, deer, rhinoceros and shrews. In the middle levels, certain species of monkey, squirrel, frog, lizard, and tree-climbing big cats, can be found. But it's in the canopy and emergent layers that 80% of the animals live. Many of them are now endangered species.

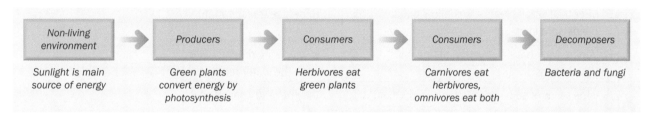

Non-living environment		Producers		Consumers		Consumers		Decomposers
Sunlight is main source of energy		Green plants convert energy by photosynthesis		Herbivores eat green plants		Carnivores eat herbivores, omnivores eat both		Bacteria and fungi

▲ Energy flows through the food chain

The tropical desert ecosystem

The characteristics of the vegetation and how it is adapted to climate and soils

Most plant adaptations are designed for survival with a minimum amount of water and long periods without it.

Desert soils cause difficulties for plants. Being either rocky or sandy, they are very porous, so water passes rapidly into them after rain. Sandy soils are mobile, so plants can easily be covered and are also loose, so plants can be up-rooted.

Desert soils are thin and contain very few plant nutrients, as very little organic matter is available to decompose into them and provide nutrients. Many desert soils are grey because salts are drawn up in solution after rain and deposited at the surface as the water evaporates. Only salt tolerant plants, such as saltbush, can grow in these saline soils.

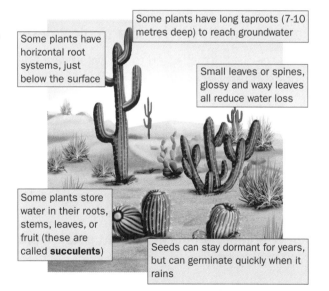

Some plants have long taproots (7-10 metres deep) to reach groundwater

Some plants have horizontal root systems, just below the surface

Small leaves or spines, glossy and waxy leaves all reduce water loss

Some plants store water in their roots, stems, leaves, or fruit (these are called **succulents**)

Seeds can stay dormant for years, but can germinate quickly when it rains

Practice questions

1. Match the following beginnings to the endings to explain how the vegetation is adapted to the environment.

	Beginnings		Endings
A	Leaves are either very small, or there are thorns, and some cacti have no thorns …	1	… to catch water after rain before it evaporates.
B	Desert vegetation is sparse. Plants are widely spaced because …	2	… after rain.
C	Most plants are low growing plants …	3	… because there is little moisture available for growth.
D	Yuccas and the small number of trees that grow have roots up to 10 metres deep …	4	… to minimise transpiration by providing shade.
E	Roots of small plants are shallow and wide-spreading …	5	… to reduce transpiration.
F	Some cacti have a covering of fine hairs on their stems …	6	… to reach the groundwater beneath the water table.
G	The seeds of some desert plants lie dormant for years. They have a very short life-cycle, flowering and producing fruit very quickly…	7	… they have to compete for water.

2. Make a table listing ways in which tropical rainforest and tropical deserts are the exact opposite of each other in their characteristics. Include climate as well as vegetation and soils. There are 20 differences in the answers but you might be able to think of more. You will have done really well if you think of more than 12.

How do animals survive in a tropical desert?

Many species have adapted to survive in very dry conditions. The zebra that migrate in the wet season into the valleys of the Namib Desert are able to detect pools of water below the surface with their nostrils. They then use their hooves to dig holes to get at the water. Some animals like elephants travel many miles from one water source to another in the Namib Desert.

In the Mojave Desert, the animals have light-coloured fur or feathers to reflect the sun. The desert tortoises feed on plants in the spring and the moisture they obtain is stored in their bladders to last them until next spring.

Example desert food chains would be:

Sun → primary producer → plant eater → predator carnivore

sun → desert grass → springbok → cheetah

sun → desert grass → springbok → lion

Because both lions and cheetah eat springbok, they would be linked on a food web.

The natural environment and human activities (1)

1 Volcanoes

Dangers from volcanoes

You should know how some of the following features present danger:

- Ash falls
- Pyroclastic flows
- Lateral blasts
- Mudflows (lahars)
- Volcanic gases
- Acid rain
- Post-eruption famine and disease
- Tsunamis
- Lava flows

Reducing the risk from volcanoes

- Lava flow diversion
- Mudflow barriers
- Building design
- Volcano monitoring: small earthquakes, ground deformation and gas emissions
- Remote sensing
- Hazard mapping and planning

Advantages brought by volcanoes

- Geothermal power
- Fertile soils
- Enlarging land area
- Tourism and associated employment
- Minerals and mining

Practice questions

1. Match the terms with their definitions.

	Term		Definition
A	Lateral blast	1	A volcano which has not erupted for some time but may erupt in the future.
B	Epicentre	2	A mudflow which contains material from volcanic eruptions.
C	Tsunami	3	When a volcano erupts sideways with great force producing gas and pyroclastic material.
D	Extinct	4	A volcano which will never erupt again.
E	Dormant	5	A powerful current of water mixed with mud, rocks and boulders running down a river channel, valley or slope.
F	Lahar	6	An ocean wave produced when there is movement of the sea bed by the fault movement which causes an earthquake. This could also be caused by the collapse of a volcanic cone into the sea.
G	Mudflow	7	Refers to the solid material produced during a violent volcanic eruption.
H	Pyroclastic	8	The point on the Earth's surface directly above an earthquake focus.

2. Choose three methods of reducing the risk from volcanoes and explain the methods you have chosen.

3. Choose three advantages brought by volcanoes and explain the advantages you have chosen.

2 Earthquakes

The effects of an earthquake are described on a 12 point scale called the Mercalli Scale. (See Chapter 11)

Effects of earthquakes

- Walls crack
- Buildings collapse
- Trees fall
- The ground cracks
- Gas pipes break causing fires
- Water and sewerage pipes break causing water shortages and disease
- Landslides
- Tsunamis

Reducing the risk from earthquakes

The ability of an area to recover from a major earthquake is affected by how wealthy a country is and how efficient the government is.

In MEDCs like Japan:

- Buildings are constructed to withstand earthquakes
- Schools and other public buildings have regular earthquake drills so that people are prepared when an earthquake strikes
- Sea walls to defend against tsunami
- Some countries are part of an international tsunami warning system

In some LEDCs buildings are not constructed to withstand a major earthquake. People are housed in crowded conditions. Government is not well organised and the countries are poor.

Practice question

1. Choose two examples of major earthquakes, one from an LEDC and one from an MEDC. For each earthquake, describe the effects of the earthquake and how the country recovered.

3 Tropical storms

When and where do tropical storms develop?

They form over oceans between May and November in the Northern Hemisphere and between November and May in the Southern Hemisphere.

Key
9	Average number of hurricanes per year
Cyclones	Local name
	Sea temperature over 27°C

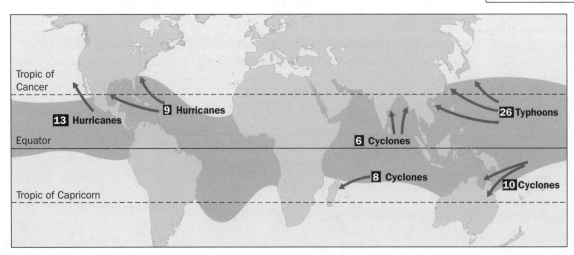

		... before the eye	... in the eye	... after the eye
Weather	**Wind strength**	Gusty then increases rapidly near the eye	Calm	Very strong again
	Wind direction	Opposite to after the eye	No wind	Opposite to before the eye
	Cloud	Heavy cumulonimbus	No cloud-sunny	Heavy cumulonimbus
	Precipitation	Torrential rain starts suddenly with thunderstorms	Dry	Intense rainfall, but less than before the eye

As in the case of earthquakes, people in LEDCs suffer more as a result of tropical storms than people in MEDCs, because they are less well-prepared.

Effects of tropical storms

- Loss of life
- Strong winds destroy buildings, even concrete and steel structures
- Power and communication lines are destroyed
- Schools, hospitals, government buildings, banks and other important buildings are destroyed
- Heavy rainfall causes flooding
- Heavy rainfall causes landslides and mudflows
- Winds whip up massive waves which cause much damage to coastal areas
- Trees are uprooted
- Crops are destroyed
- Ships at sea are wrecked and sunk
- Effects are greatest over the heavily populated areas
- Costs millions of $

Reducing the risk from tropical storms

- Satellite images to track storms to provide warning (although predictions of where the storm will hit are not always successful. Tropical storms can suddenly change course and when the storm does come, there will have been a considerable economic cost and disruption for nothing)
- Evacuation plans for the population
- Sea walls and artificial levées to prevent flooding
- Buildings built to withstand strong winds
- Emergency supplies of food and water stored in advance of the storm
- Extra toilets
- Put strong covers over windows and evacuate
- Emergency shelters, well stocked with food and water
- Insure properties

Practice questions

1. Describe the features of a tropical storm that cause destruction and loss of life.

2. Choose an example of a tropical storm. Describe the features of the weather, the effects of the storm and the response of the people and the government.

4 River valleys and flooding

Advantages of living in river valleys

- Flat land makes it easy to build roads and settlements and carry out agricultural practices
- Soils are often mineral-rich and fertile due to the silt and mud deposited by the river during floods (alluvium), so agriculture is profitable
- Valleys are often natural routeways
- Rivers may be navigable allowing transport and trade
- Rivers may contain fish as a food supply
- Rivers provide water supply
- Features such as waterfalls may be tourist attractions

Causes of flooding

- Heavy, continuous rainfall on already saturated ground
- Steep slopes that increase the rate of runoff
- Impermeable bedrock that rain cannot soak into
- Urbanisation, which creates impermeable surfaces
- Deforestation, which increases runoff

Effects of flooding

You should be able to describe the different effects of floods in LEDC and MEDC examples.

Flood prevention and protection

- **Planting vegetation** such as trees which take in rain water through their roots and lose it by transpiration, acting like a sponge, so that it does not reach the river
- Water can be trapped in **reservoirs** on the tributaries, then released slowly over a longer period of time
- **Straightening the channel** shortens the river and moves the water away faster
- **Dredging the channel** makes it deeper and increases its capacity
- **Artificial levées** increase the capacity of the channel, just like dredging
- **Bridge design** – Modern bridges are slim and streamlined to prevent them acting like dams
- **Wash lands** – Flood waters can be channelled into areas where the damage will be less

Practice questions

1. Choose an example of a river valley.
   a) Explain why it has a dense population.
   b) Describe the causes of flooding in the area you have chosen.
   c) Describe the consequences of the flooding for the people of the area.
   d) Explain what has been done to help the people or to prevent future floods.

2. With reference to examples, explain how floods have different effects in MEDCs and in LEDCs.

The natural environment and human activities (2)

1 Drought

> ### KEY IDEAS
>
> → **Drought** is a longer than usual period of dry weather.
> → Many people live in areas where the climate has a dry season and some live in deserts where it is dry all year but they adjust to those conditions and learn to cope with the difficulties of the environment.
> → Droughts occur when rain fails at a time when it is expected to fall, so they cause problems for vegetation and human activities. The effects are worse in LEDCs but they do affect MEDCs such as Australia.

Where droughts occur

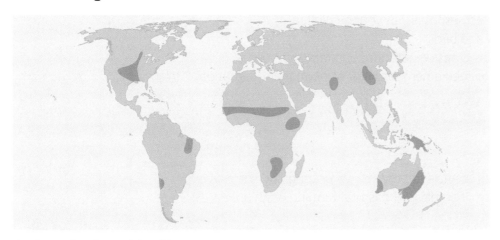

▲ Areas where frequent drought occurs

In some of these areas the effect of drought on the vegetation and farmlands has been so severe that they have experienced **desertification**; they became dry and like a desert.

Why drought occurs

In semi-desert areas, rainfall in a year is often very different from the mean rainfall. Drought years can be followed by wet years. The southern fringe of the Sahara Desert is known as the Sahel. Here the Sahara Desert was thought to be advancing southwards during severe droughts between 1970 and 1993.

As rain falls in the summer months, high temperatures evaporate some of it before it can sink into the soil or reach the **water table**.

How people can make the drought problem worse

During wet years people keep more animals. When there is a drought there is then **overstocking** of animals – too many animals for the pasture available. **Overgrazing** then removes all the pasture. The soil is then exposed and, with no roots to hold it, is easily blown away.

This **soil erosion** is also assisted by the dryness of the soil. Dry soil is lighter than wet soil and more easily removed. In addition, increased grazing results in less **humus** in the soil, causing a breakdown in its structure so that it crumbles into individual particles which are lighter and easily blown away.

Population growth leads to trees and bushes being removed for firewood, again leaving the soil more exposed to wind erosion. The higher population takes more water from the boreholes for themselves and their families.

Overcultivation to provide food for the growing population leads to **soil exhaustion**, by which the nutrients are removed until all fertility is lost.

Consequences of drought

- Crop failure and reduced crop yields, leading to rising world food prices for produce such as wheat
- Insufficient food and water for the people and their animals
- Pastures die, leading to death of livestock
- Water supplies dry up, leading to conflicts between agricultural and domestic use
- Soil erosion
- Dust storms
- Malnutrition and sickness, especially in young children
- Death of wild animals and birds
- Bush fires which destroy crops, vegetation, habitats and homes
- Population migration to seek emergency aid, or permanent migration to towns and cities

Sustainable farming

This can involve:

- reducing the number of animals to prevent overgrazing and soil erosion
- having a better balance of arable and pastoral farming, with animal manure as a fertiliser
- planting trees to prevent soil erosion
- building earth dams to trap water during wet seasons to be used for irrigation
- sink more boreholes for water supply and install reliable pumps
- provide firewood to prevent existing trees and shrubs from being cut down

The work of aid agencies

Aid agencies carry out tasks such as:

- providing emergency medical aid, tents, food and water
- giving schools food rations
- drilling boreholes and wells
- building dams and laying water pipelines
- providing emergency farming kits, such as a drip irrigation systems and water tanks.

Aid agencies are hampered by:

- lawlessness and civil war
- the sheer scale of the problem, involving many thousands of people
- refugees from other areas coming to the aid camps.

Practice questions

1.

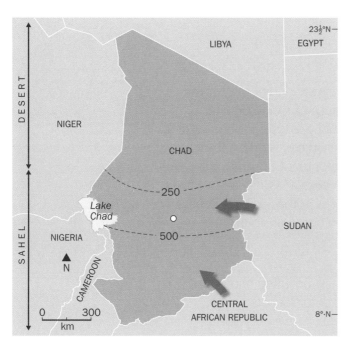

a) What is an isohyet?

b) Suggest why the refugees from other countries have moved to the south of Chad and not the north.

c) Suggest why droughts occur in Chad.

2. Look at the table of rainfall in part of south-east Australia. In which years were drought and flooding likely to be problems? Explain your answer.

Year	Difference from the long-term mean (in mm)	Year	Difference from the long-term mean (in mm)
1995	+44	2003	−50
1996	+52	2004	−21
1997	−45	2005	−46
1998	+70	2006	−206
1999	+79	2007	−10
2000	+68	2008	−37
2001	−90	2009	−78
2002	−208	2010	+320

▲ Differences between annual rainfalls (1995–2010) and the mean rainfall in the Murray River Basin of south-east Australia

3. Choose an example of drought that you have studied. Describe the consequences of drought in the example you have chosen.

2 Tropical desert environments

Refer to the map on page 91 in Chapter 21.

Development in the deserts

- Irrigated agriculture using exotic rivers such as the Nile or Orange
- Irrigated agriculture using boreholes and small dams
- Using exotic rivers for HEP
- Mining, for example uranium and diamonds in the Namib desert
- Tourism, including ecotourism, because of sunny climate, dramatic landscapes and wildlife. For example, Namibia, Lanzarote
- Development of urban areas, like Las Vegas, Pheonix, and Palm Springs in the Mojave Desert, often based on tourism

Problems of development in the deserts

- Over-use of water resources, including progressive lowering of the water table, so that supplies may dry up without the development of re-cycling
- Conflicts over water use, for example, between tourism and agriculture
- Damage to delicate vegetation by mining and tourism
- Visually polluting mines that may deter tourists
- Rainfall is very irregular, making small scale farming difficult to plan
- Flash floods destroy roads, buildings, and crops
- Tourists can create noise pollution, litter, and over-use of water

Practice questions

1. Describe the climate of Tamanrasset in Algeria which is shown by the graph.

2. Choose an example of a desert that you have studied and describe how it has been used by humans. What problems has the use of the desert created?

3. Explain the meaning of the terms *exotic river* and *oasis*.

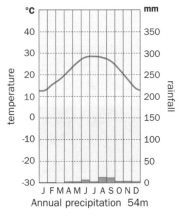

Annual precipitation 54m

3 Tropical rainforest environments

Refer to the map on page 86 in Chapter 21.

The value of tropical rainforests in maintaining the environment

- Keeping the small amount of soil fertile
- Reducing soil erosion and mudflows
- Stopping rivers being polluted by eroded soil
- Transpiration providing the necessary water for rainfall and a healthy ecosystem
- Photosynthesis taking in carbon dioxide, one of the main **greenhouse gases**, reducing **enhanced global warming**
- Absorbing solar radiation, reducing heating of the atmosphere

The value of tropical rainforests for human activities

- Diversity of species, many of which may have value as yet unknown, including medicinal plants such as quinine
- Home to many tribes
- Hardwood timber for construction and furniture
- Wood for fuel for local people
- Ecotourism - tourists interested in nature and wildlife

Loss of tropical rainforest

- Source of wealth which multi-national companies exploit for raw materials
- Establishment of commercial farming, such as cattle ranching in Brazil
- Population pressure and poverty encourage felling for farming land
- Establishment of plantations, such as oil palm (often for biodiesel) and rubber
- Establishment of hydro-electric power stations, for example, Bakun scheme in Sarawak
- Development of mining
- Roads built to access the mines, HEP sites and logging areas open up areas alongside them for deforestation and development
- Vast areas which are too large for protection agencies to patrol
- Often logging is done without re-planting

Consequences of deforestation

- Clearance of vast areas of the forest by burning has covered far away countries in smoke and ash, which is a health hazard
- Global warming - burning forests emits a lot of carbon dioxide into the atmosphere which contributes to enhanced global warming. When they are removed, less carbon dioxide is taken in and less oxygen is given out. Forests are needed to combat the effects of the extra carbon dioxide in the atmosphere from burning fossil fuels
- Reduction in rainfall and possible drought because of reduced transpiration
- Loss of biodiversity - loss of plant species which may have unknown uses as medicines, industrial raw materials and foods
- End of ways of life and unique cultures, such as hunter-gatherers and shifting cultivation
- Loss of habitats for animals, such as the orang-utan in Borneo.
- Loss of soil fertility and soil erosion - bare soils lose fertility and become eroded, preventing further agricultural use
- Rivers are choked with eroded soil
- As soils are degraded as a result of deforestation, any forest which is allowed to regenerate (secondary growth) is always poorer than the original forest

Managing tropical rainforests

- Creating National Parks to protect wildlife and habitats
- Afforestation and reforestation
- Developing ecotourism
- Development and use of more efficient farming methods
- Educating the people about why forest conservation is needed

Theme 2: The natural environment

Exam-style questions

1. a)

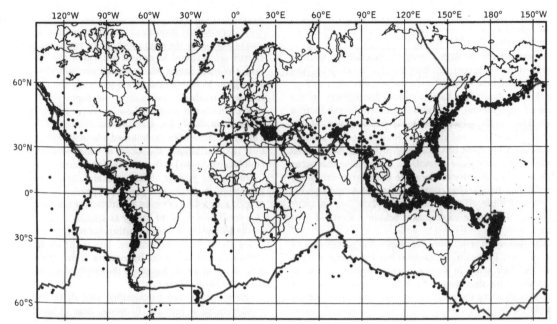

i) Define the term earthquake *epicentre*. [1]

ii) Look at the map of the world showing the locations of earthquake epicentres. Name two other features found in these zones. [2]

iii) Describe the distribution of earthquakes shown on the map. [3]

iv) Explain the distribution shown on the map. [4]

b) The map gives information about the intensity (strength) of an earthquake in California USA, in 1994.

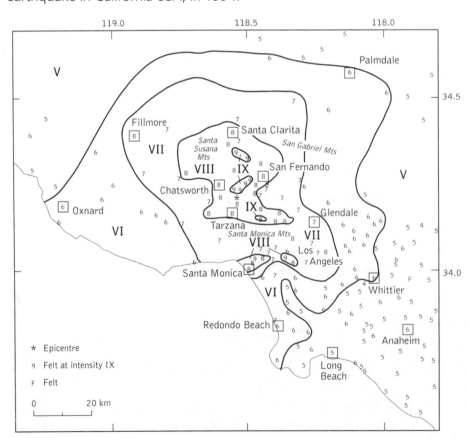

     i) Describe the pattern of intensity of the earthquake. [3]
     ii) Describe the effects of a severe earthquake. [5]

  c) Choose a major earthquake that you have studied. Explain why it occurred at that location and describe how people responded to the earthquake. [7]

2. a) Study the diagram which shows two river valleys.

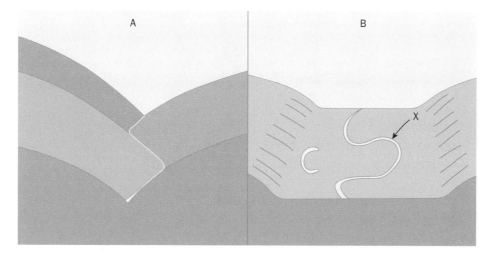

     i) Name the feature of the river at X on the diagram. [1]
     ii) Which part of the river valley do the diagrams show, the *upper course*, the *middle course* or the *lower course*? [2]
     iii) Using the diagram, name one feature of river valley A and two features of river valley B. [3]
     iv) Describe four processes of river erosion. [4]

  b)

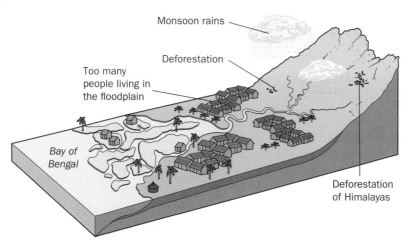

     i) Choose three features shown on the diagram of Bangladesh and explain how they lead to deaths due to flooding [3]
     ii) Explain what can be done to reduce the effects of river floods. [5]

  c) For a major river valley that you have studied, explain the advantages and disadvantages of living there. [7]

3. Look at the profile of part of a coast.

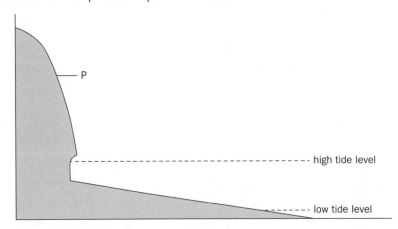

high tide level

low tide level

a)

    i) Identify the feature labelled P. [1]

    ii) Using the diagram only, describe the wave-cut platform. [3]

    iii) Using your own knowledge, describe **other** features of wave-cut platforms. [3]

    iv) Describe the stages in the formation of a wave-cut platform. [5]

b) Name and describe three processes of marine erosion which may have helped to erode the wave-cut platform. [6]

c) Name a stretch of coast you have studied and describe the features of coastal deposition along it. [7]

4. a) Study the tropical desert climate graph for Cairo at 30° North. Use the graph to answer the following questions.

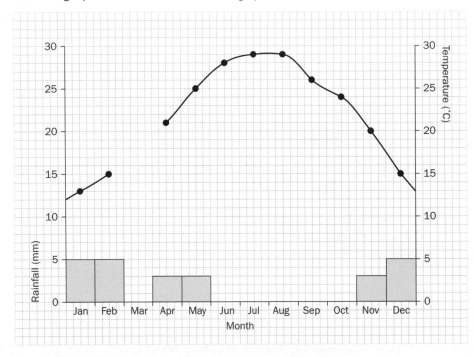

    i) Complete the graph by plotting the data for March. [2]

	Temperature (°C)	Precipitation (mm)
March	18	5

ii) Calculate the annual temperature range. Show your working. [2]
iii) Describe the annual rainfall total and the rainfall distribution at Cairo. [3]
iv) Describe the temperature characteristics of Cairo. [4]

b) Explain why tropical deserts can have shade temperatures as high as 50°C during the day and as low as 6°C at night. [3]

c) Study the rainfall data, taken over a period of 34 years, for Aswan (24°N), a place that climate graphs show as having no annual rainfall.

Mean annual rainfall (mm)	Maximum rainfall in a year (mm)	Minimum rainfall in a year (mm)	Maximum rainfall in 24 hours (mm)
0.22	27.2	0	8.6

Using the information given, describe the rainfall characteristics of Aswan. [4]

d) Describe the natural vegetation of a named tropical desert that you have studied and explain how it is adapted to survive in the difficult climatic conditions. [7]

Agriculture and large-scale commercial farming

Agriculture is farming. It is the artificial cultivation of plants (crops) and rearing of animals for food and other products. Agriculture can involve crops (arable farming), the rearing of animals (pastoral farming) or both (mixed farming).

Agriculture can be described as a system because it has **inputs**, **processes** and **outputs**.

Physical inputs These are the provided by nature	Human inputs These are provided by people
Climate: temperatures, rainfall, sunshine Soil Land and its relief	Capital Labour Machinery and tools Seeds Social structures Government influence Market influence Fertilisers, pesticides and herbicides Irrigation

Processes

These are the methods people use to produce the outputs:

- Preparation of the land: clearing vegetation, providing terracing, drainage and irrigation systems
- Ploughing
- Sowing
- Weeding
- Application of fertilisers, pesticides, herbicides and irrigation
- Harvesting
- Storage and transporting to market

Outputs

These are the products of the system:

- Crops
- Meat
- Milk
- Industrial products such as cotton, rubber or leather

Commercial farming

The farmer sells his or her output to make a profit. This is typical of modern, large-scale farming. The crops produced are known as **cash crops** – they are sold for money.

Subsistence farming

The farmer grows crops or rears animals for consumption for his or her family. The crops are called **subsistence crops**. A surplus may be produced from time to time which is sold.

	Commercial farming	Subsistence farming
Capital (money)	Large capital input, sometimes from international companies	A complete lack of capital may prevent any increase in output
Land	Large area	Very small farms
Labour	Paid labour (often skilled), much use of research and development	Family labour, relying on traditional methods
Machinery and tools	Much use of mechanisation for all processes	Hand tools, such as hoes, and ploughs sometimes pulled by draught animals
Seeds	Improved varieties and hybrids	Seeds left over from the previous year's crop
Market influence	Production is geared to current market demands and prices	No market influence
Fertilisers	Generally used	Used much less, although sometimes animal manure is available
Pesticides and herbicides	Generally used	Used much less
Irrigation in dry areas	Uses complex systems	Either none or very low-technology systems

▲ Comparing commercial and subsistence farming

Intensive and extensive farming

	An intensive farm	An extensive farm
Area of land	Small	Large
Large machines	Few	Many
Labour input per hectare	High	Low
Fertiliser input per hectare	High	Low
Output per hectare	High	Low

▲ Comparing intensive and extensive farms

These terms can apply to either commercial or subsistence agriculture. Commercial farms may be intensive or extensive, as may subsistence farms, although the latter tend to be intensive.

As the table shows, on an extensive farm the inputs per hectare are low and the outputs per hectare are low, but this is overcome by using a large area of land.

Large-scale system of commercial farming

Physical inputs	Human inputs
Very large area of land. The farm may be run on extensive principles but not always. You should be able to describe specific physical inputs (climate, soils and relief) of examples which you have studied.	Large capital input. In some cases the farm may be backed by a multi-national corporation. A paid, often skilled labour force. Much use of research and development. Modern mechanisation. Improved crop varieties and hybrids. Fertiliser, pesticide, herbicide. Irrigation where necessary. Strong market links.

Practice questions

1. Which of the following are physical inputs, human inputs, processes and outputs? Place one tick on each row.

	Physical input	Human input	Process	Output
Labour				
Milk				
Rainfall,				
Ploughing				
Soil				
Storage and transporting to market				
Capital				
Pesticides				
Application of fertilisers, pesticides, herbicides and irrigation				
Government influence				
Irrigation				
Fertilisers				
Harvesting				
Herbicides				
Temperatures				
Drainage				
Sowing				
Land				
Market influence				
Weeding				
Machinery				
Seeds				
Meat				
Sunshine				
Products such as cotton, rubber or leather				
Social structures				

2. Explain the differences between intensive and extensive farming.

3. Choose an example of large-scale, commercial farming which you have studies. Write an account of the farming under the following headings: physical inputs, human inputs, methods, outputs. Remember to give specific details of the location that you have chosen.

4. Complete the following crossword.

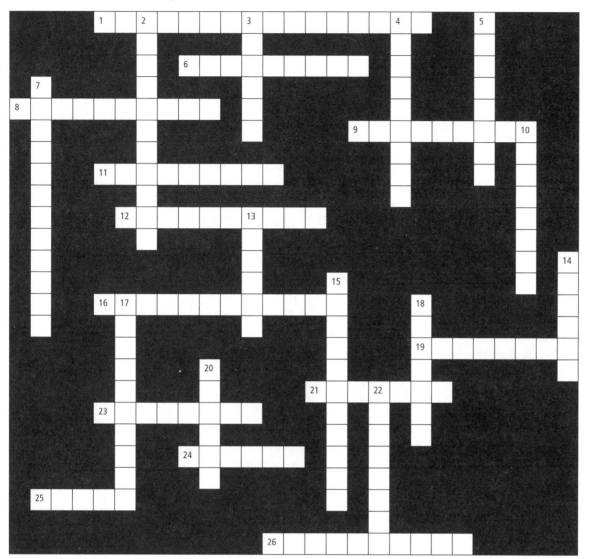

Across	Down
1 A machine used by cereal famers. It carries out different processes. (7/9)	2 Growing the same crop on the same plot year after year. (11)
6 Weedkiller. (9)	3 A crop variety produced by cross-breeding existing varieties. (6)
8 Farming to make a surplus to sell for profit. (10)	4 Farming without achieving the maximum output from each plot of land but compensating for this by having a large area. (9)
9 These chemical are used to kill insects which affect crops or animals. (9)	5 Mineral-rich, fertile soils deposited by a river. (8)
11 Farming to achieve the maximum output from each plot of land. (9)	7 The disease that affects sheep and cattle and is controlled by quarantines and occasionally the elimination of millions of animals. (4/3/5)
12 Farming methods used to stop soil erosion in areas of low rainfall. (3/7)	10 Storage towers used to store grain, especially in Canada. (8)
16 Farming with both crops and animals. (5/7)	13 Animal dung or vegetable waste added to soil to increase its fertility. (6)
19 An input provided by nature such as land, rainfall and soil. (8)	14 A farming system involving growing crops. (6)
21 The products of the farming system. (7)	15 Farming. (11)
23 This crop is to be sold. (4/4)	17 Providing crops with water artificially. (10)
24 The place where there is a demand for the farming products. (6)	18 The money invested in the farm. (7)
25 An input provided for agriculture with assistance from people such as labour or machinery. (5)	20 A grain crop like wheat, maize or millet. (6)
26 Chemicals added to the soil to make it more productive. (10)	22 A farming system involving keeping animals. (8)

28 Small-scale subsistence farming and food shortages

Natural inputs of subsistence farming

In arable systems, each farm is small, perhaps as small as 1–3 ha. In the case of subsistence pastoral farming, areas of land can be much larger, especially in the case of pastoral nomadism, where people move from place to place with their animals.

Human inputs and processes of subsistence farming

- The processes are **intensive** (see Chapter 27).
- **Capital** input is small which prevents many subsistence farmers from increasing their output. Family **labour** is generally used.
- **Tradition** often fixes the roles of men and women in different ways in different societies. Hand **tools** such as hoes are used and draught animals such as oxen or water buffalo are used to pull ploughs. There are few **machines**.
- **Seeds** left over from the previous year's crop are used for the next year which prevents the use of improved varieties.
- The only **fertiliser** used might be animal **manure**, although in many areas this is used as a fuel, preventing soil improvement.
- Where **irrigation** is used, very low technology systems are in place, usually draining water in channels from a nearby stream.

Shifting cultivation

Today this system is only practised in a few areas. An area of land is cleared, and ash from burnt vegetation is used as fertiliser. The land is cultivated for a few years in the traditional manner until it is exhausted and crop yields decline. The people then move to another area, sometimes building a new settlement, and repeat the process, not returning to the original plot for perhaps 20 years. This system is still used in some tropical areas where the soil fertility is low and minerals are leached by heavy rainfall.

Food shortages

Food shortages are a particular problem in many LEDCs of Africa and south Asia.

It is important that world food supply grows to keep pace with the growing world population.

Globally, more than one third of child deaths are attributable to under-nutrition. Undernourished children have lowered resistance to infection and are more likely to die from common childhood ailments like diarrhoea and respiratory infections. There are particular diseases linked to protein deficiency e.g. marasmus and kwashiorkor.

Well-nourished women face fewer risks during pregnancy and childbirth, and their children get a better start in life.

Causes of food shortages

These are linked to crop failure and poor agricultural yields.

- **Soil exhaustion** as a result of overcropping, monoculture, insufficient fertiliser and manure.
- **Drought**, particularly in areas of the tropics where seasonal rainfall occurs.
- **Floods.** Where farming occurs on flood plains serious flooding can lead to the complete loss of a year's harvest.
- **Tropical cyclones.** Crops can be destroyed by strong winds, torrential rain or the associated floods.
- **Pests** include locusts and birds which destroy mature crops.
- **Crop diseases** can destroy crops in the fields or during storage.
- **Animal diseases** such as foot and mouth disease, nagana or trypanosomiasis result in low production of meat and milk or death of animals.
- **Diseases affecting** farmers, such as malaria, HIV/AIDS.
- **Low capital investment** Many people who practise subsistence agriculture are stuck in a vicious circle of poverty, (see the diagram on this page).
- **Poor transport**. Farmers in remote areas find it difficult to sell surpluses and raise capital, receive supplies and information about possible improvements to farming.
- **Wars**. Where people are forced to leave their homes and become refugees and have uncertain futures has an obvious effect on their ability to make long term investments in increased food production.
- **Increased use of biofuels.** From 2008 to 2011 some land previously used for the production of food was changed to produce crops for biofuel production.

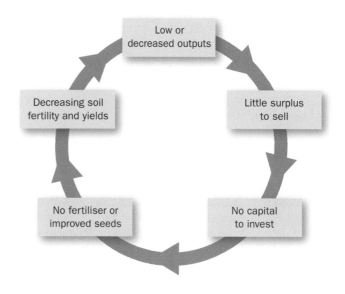

Solutions to the problem of food shortages and improving subsistence agriculture

- New **hybrid seed varieties** can be very responsive to fertilisers, give higher yields and have shorter growing seasons (although you may now that these may bring other problems). Genetically modified (GM) crops can have the same effects.
- **Extend irrigation in dry areas.**
- **Different crops** for example alternatives to maize include sorghum, sweet potatoes, cassava and groundnuts.
- **Subsidised farming inputs** such as tractors, seeds and fertiliser, help overcome the lack of capital.
- **Disease control**, for example methods to control foot and mouth disease such as fencing.
- **Education and training of farmers** in new methods or growing different crops.
- **Improved markets** for crops to stimulate production.
- **Measures to control soil erosion**, including inter-cropping, terracing, contour ploughing, crop rotation and reducing stocking densities.

Practice questions

1. Name an example of small-scale subsistence farming. For the example you have chosen, write an account of the physical inputs, human inputs, processes and outputs.

2. Explain the main causes of food shortages and what might be done to increase agricultural output on subsistence farms.

3. Match the terms with their definitions

	Term		Definition
A	Fertiliser	1	Method of preventing soil erosion. Crops are grown in narrow bands, often at right angles to the prevailing wind, with other bands of different crops in between. The crops are harvested at different times so the field is never completely bare.
B	Overcropping	2	Changing the crop on a plot every year for three or four years before the first crop is grown again.
C	Pastoral	3	Animals used to pull tools or machines such as ploughs.
D	Soil exhaustion	4	A disease of the human immune system caused by the human immunodeficiency virus.
E	Manure	5	Providing crops with water by artificial means, i.e. not directly from rainfall.
F	Foot and mouth disease	6	An infectious and sometimes fatal disease that affects sheep and cattle and some wild species.
G	HIV/AIDS	7	Growing the same crop on the same plot repeatedly.
H	Crop rotation	8	An improved variety produced by cross-breeding existing varieties.
I	Hybrid	9	The removal of the soil by wind or running water on slopes.
J	Intensive farming	10	Farming to achieve the maximum output from each plot of land.
K	Draught animals	11	Ploughing around a hill to prevent soil erosion.
L	Irrigation	12	A form of severe protein malnutrition, especially in children after weaning, marked by lethargy, growth retardation, anaemia, potbelly, skin depigmentation, and hair loss or change in hair colour.
M	Kwashiorkor	13	A potentially fatal tropical disease spread through the bite of an infected female mosquito.
N	Contour ploughing	14	A system of arable agriculture where there is no permanent home and the plots are cultivated for two or three years then abandoned for many years.
O	Marasmus	15	Chemicals added to the soil to make it more productive.
P	Overgrazing	16	When minerals in the soil have become depleted due to overcropping, resulting in decreasing crop yields.
Q	Nagana	17	Growing crops too close together or too frequently on a plot so that the soil loses fertility.
R	Shifting cultivation	18	A form of severe protein malnutrition characterized by a lack of energy. It particularly affects children up to the age of one. Body weight may be reduced to less than 80% of the average weight for the height.
S	Soil erosion	19	Grazing too many animals on a plot of land with the result that the quality of the pasture deteriorates.
T	Strip cultivation	20	A system used to grow crops on sloping land by creating flat steps.
U	Malaria	21	Keeping animals.
V	Subsistence farming	22	Farming to feed oneself and one's family.
W	Monoculture	23	Any organic matter added to the soil to improve its structure and mineral content, e.g. animal dung or vegetable waste.
X	Terracing	24	A disease affecting cattle spread by the tsetse fly. Called sleeping sickness in humans.

4.

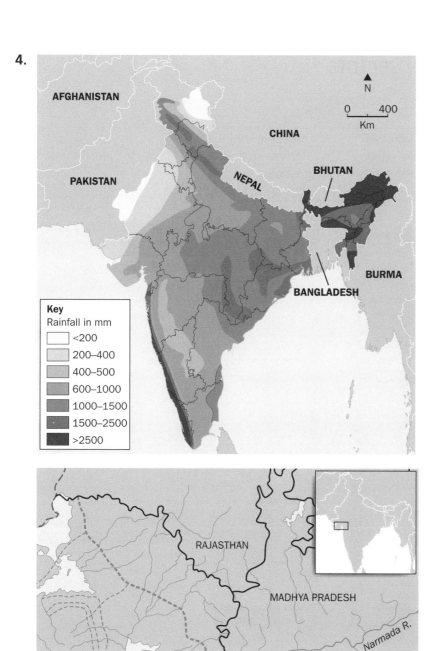

a) Which area of India will have the greatest need for irrigation water and why?

b) The map shows the location of the Sardar Sarovar Dam on the Narmada River in India. As well as to supply drinking water and electricity, the dam is to supply irrigation water to dry areas. What is the annual rainfall in the Narmada River area?

c) Look at the canal that leads from the Sardar Sarovar Dam. In which compass direction does the canal go and why?

Industrial systems

Industrial sectors

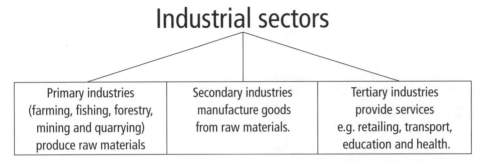

| Primary industries (farming, fishing, forestry, mining and quarrying) produce raw materials | Secondary industries manufacture goods from raw materials. | Tertiary industries provide services e.g. retailing, transport, education and health. |

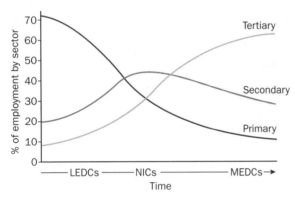

Employment in industrial sectors changes with time as a country becomes more economically developed.

Industrial systems

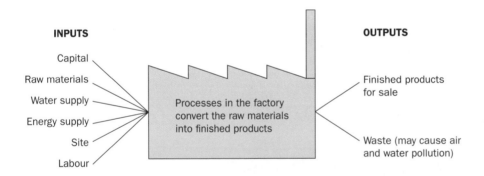

The influence of inputs on the processes and outputs of industrial systems

Some industries have inputs which require a lot of processing and can lead to much pollution.

Industry	Inputs	Processes	Outputs	Possible adverse results
iron and steel manufacturing	iron ore, coke, limestone to separate the iron from impurities in the ore, water, recycled scrap iron. For special steel: alloys, e.g. chromium, cobalt	heating of ore to separate the iron by burning coke, rolling into sheets, cutting into lengths	cast iron and pig iron, waste: slag, gases (sulfur dioxide, carbon dioxide, nitrous and nitric oxides, hydrogen sulfide)	noise, large ugly buildings, slag heaps, dust, air pollution, water pollution (contaminated cooling water and scrubber effluent), risk of fires and explosions

Factors affecting industrial location

Physical factors

Raw materials	Raw materials for heavy industries are bulky and expensive to transport. Nearness to iron ore or coal mines or, if imported raw materials were used, to importing ports used to be the most important factor in their location. Today's transport systems are more economical and bulky commodities are transported across the globe in large ships. Market and government influence have become more important now in the locations of vehicle manufacture. Light industries use materials and have products which are small in volume but of very high value so transport costs are of little importance.
Site	This is important for large-scale manufacturing, such as oil refining, or chemical manufacture. The large size of the plant means that very large areas of flat land are needed. It also needs to be well-drained and on solid bedrock.
Energy	Energy supplies were important in the past when sites next to fast flowing rivers or coal mines were favoured. Today, a link to the electricity grid is often sufficient. For a few manufacturing industries, e.g. aluminium smelting, which require very large amounts of power, access to cheap energy is important.
Water supply	Some industries, such as paper, chemicals and metals, require water in high quantities and may need to be located where they can have their own supplies from rivers or boreholes.
Natural harbours and route centres	Ports are favoured locations because raw materials can be imported and products exported at less cost. Major roads and railways often follow natural routes such as valleys. Industries locate along them and at junctions of routes.

Human and economic factors

Capital	The start-up costs of an industry may come from other businesses, banks or governments. Finance may be more freely available in some countries or areas than in others. It is often connected with political factors (see below). Only global trans-national companies have access to the capital needed for the motor vehicle industry which achieves economies of scale by mass production on very large assembly lines using complex machinery.
Labour	The size of the labour force is important in many industries where thousands are employed. Usually, the quality of the labour force is just as important as its size. Workers also need to be adaptable to change. The reputation of the workforce in an area is very important today. Large numbers of skilled workers are needed for vehicle manufacturing to operate highly complex production lines, so the company seeks educated, hard-working employees in countries with few restrictive employment laws. Workers in LEDCs are about 40% cheaper than in MEDCs.
Transport	Transport is still an important factor in the location of industries producing bulky goods. For vehicle manufacturing, access to ports and good road systems is important for the assembly of components and distribution of vehicles to markets. The massive increase in the use of containers, which can carry many items securely and cheaply, has helped the development of industries in Asia's NICs. Trans-national companies are able to take advantage of the low production costs (cheaper wages) there and to use containers to transport the products to the rich markets in MEDCs.
Markets	This is the most important factor in the location of the motor vehicle industry today; there has been rapid expansion into NICs with very large populations, such as China, India, Brazil and Indonesia.
Political influence	Governments influence the location of industry by providing financial incentives to companies to locate in particular areas. The tax system of a country is also an important influence on decisions taken by trans-national companies. As many countries have higher taxation levels on imported manufactured goods than on components, it is cheaper to transport the components for motor vehicles and assemble them in a country where there is a large market, than to export cars to the country.
Quality of life	This influences the location of industries which require a highly-skilled professional workforce that prefers pleasant areas with good housing and leisure facilities.

High technology industry

Processes

- The industry has a high degree of research and development work to keep ahead of competitors.
- The manufacturing is highly automated and computerised, with the most advanced technology used to make the products.

Outputs

Products include pharmaceuticals, precision instruments, computers, televisions, mobile phones and aircraft. Biotechnology companies develop new kinds of food, drink and vaccines.

Inputs and their effects on the location of high technology industry

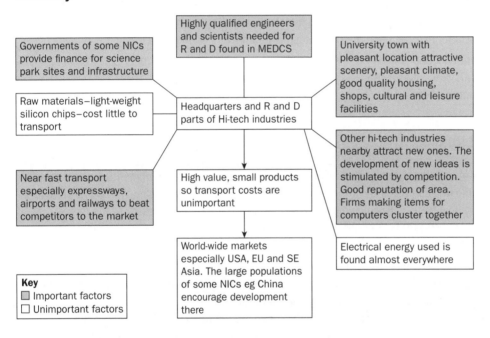

Practice questions

1. Use the information about the steelworks at Anshan to list reasons why the steelworks has developed there. Put the factors in rank order.

Anshan, a city in China with 3.65 million people, has a steelworks on a large area of flat land. It has been modernised with capital from the government. Local iron ore and limestone are heated together in the blast furnace with coke used as the fuel. The coke is made from coal transported by rail from Benxi, 130 kilometres away. Limestone helps to separate impurities from the iron and they are drained away as waste material called slag. Pig iron is changed into steel by burning out most of the remaining impurities in an electric furnace. Oxygen is blown into it and carbon mixes with it, so carbon dioxide is released. The steel is rolled and cut into sheets, bars, pipes and other shapes used in engineering, shipbuilding, motor vehicle manufacturing and other industries which have grown in the area because of the availability of iron and steel. They provide a large market for the steel.

2. Explain why
   a) the Headquarters and Research and Development parts of high technology firms are mainly in MEDCs.
   b) branch factories undertaking assembly of the products are sometimes in LEDCs.

3. Toyota manufactures and sells vehicles in 170 countries.
   a) Why is this an advantage?
   b) What makes this 'globalisation' possible?

4. Suggest why
   a) motor vehicle manufacturing needs a large, flat site,
   b) factories are often located away from higher class residential areas,
   c) factories are often sited downwind of residential areas.

5. Match the terms to the definitions.

	Term		Definition
A	Globalisation	1	Services needed for economic activity.
B	Economies of scale	2	Land not previously built on.
C	Infrastructure	3	An industry that uses the most advanced technology to make products.
D	Heavy industry	4	The subdivision of industries into primary, secondary and tertiary.
E	Footloose	5	Industries dealing with raw materials and products which have little weight or bulk.
F	High technology industry	6	The things which will be converted into the finished product.
G	Assembly line	7	A method of car production where each car passes through a series of stages and each stage concentrates on one part of the production process.
H	Greenfield site	8	An industry that is not tied to any particular type of location.
I	Raw materials	9	Industries dealing with bulky raw materials and products.
J	Light industry	10	The way that one large unit can operate more cheaply and efficiently than several smaller units.
K	Industrial sectors	11	The process by which local economies, societies, and cultures have become integrated through communication, transportation and trade. Economic globalisation is the integration of national economies into the international economy through trade, foreign investment and flow of capital across international boundaries.
L	Trans-national corporation	12	Industries which provide the needs of people, other than goods. Service industries.
M	Primary industry	13	Machines and other tools.
N	Silicon chip (microchip)	14	Industries which produce raw materials.
O	Scrubber	15	A device for removing impurities from gases.
P	Secondary industry	16	Industries which convert raw materials into finished goods. Manufacturing industries.
Q	Technology	17	A company which has its operation in several countries. A multi-national company.
R	Tertiary industry	18	A small wafer of silicon or other semiconductor material on which circuits are laid out.

Economic benefits of tourism	Negative economic effects of tourism
Reduced unemployment. Employment and income raise standards of living.	Seasonal unemployment, e.g. in the hurricane season of the Caribbean.
Businesses connected with tourism, such as hotels and coach companies, make a profit and employ more staff. Local craft industries benefit as tourists buy souvenirs. Farmers sell food to hotels.	Many hotels and other businesses connected with tourism are owned by foreign companies, so much of the profits go to other countries.
Other local businesses, such as shops, benefit from the increased wealth and spending power. They take on more workers.	People move to the coasts to work in tourism, so the inland economy, e.g. farming, declines.
The employees and businesses pay taxes and increase the wealth of the country so more development of the country can be funded.	Some workers are foreigners who send their wages home.
Skills learnt while working in tourism can be used in other occupations and areas of the country.	Skills gained through jobs in tourism enable some to leave the country to work in other countries.
Tourist areas become richer with more facilities, including health and education. (Tourism provides nearly 60% of the GDP of the Maldives).	Parts of the country without tourist attractions remain poorer. Inequalities increase between them and the richer tourist resorts.
Foreign exchange increases as tourists change money into the local currency. They pay a fee to do so and the exchange rate is always favourable to the country.	In periods of economic downturn in MEDCs, businesses and workers in tourism lose income because fewer tourists travel.

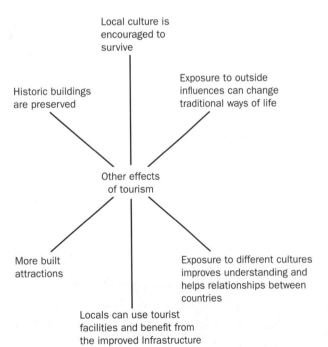

Local culture is encouraged to survive

Historic buildings are preserved

Exposure to outside influences can change traditional ways of life

Other effects of tourism

More built attractions

Exposure to different cultures improves understanding and helps relationships between countries

Locals can use tourist facilities and benefit from the improved Infrastructure

Ecotourism, which aims to allow tourists to visit while preserving the environment, and community-based tourism are becoming more popular in fragile environments of LEDCs. Both are small-scale. Community-based tourism aims to consult with and benefit the local community by giving jobs. Some locals accommodate tourists in their houses and act as guides.

Lanzarote, an economy transformed by tourism

Lanzarote, a small island off the coast of North Africa, has barren, volcanic, desert land without any natural surface water. It had a very poor economy.

In the 1960s, it was decided to promote tourism, so the necessary infrastructure was developed:

- good roads were made over the island
- the airport was improved
- desalinisation plants were built to convert seawater to usable water
- artificial beaches were made using sand from the Sahara Desert
- hotels and apartments were built (with a maximum height of two storeys, so as not to intrude on the natural landscape).

As much use as possible was made of the natural landscape to develop attractions.

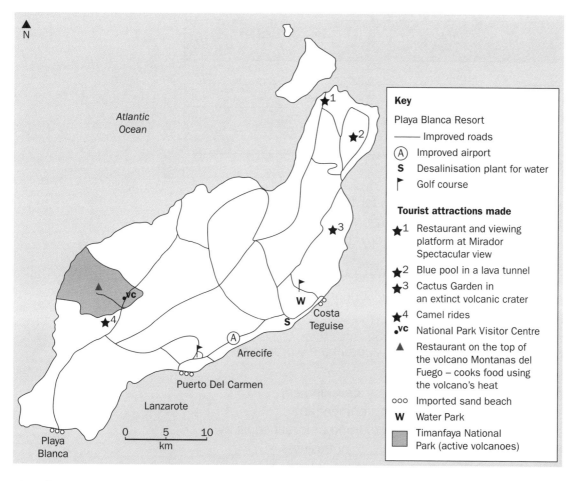

The climate makes Lanzarote an all-year round tourist destination.

- It has little rain and cloud.
- Winters are warm with guaranteed sunshine.
- Summers are hot and dry.
- The constant Trade winds have a cooling effect in the summer heat.

Most inhabitants think that enough of the landscape has been covered and that no further development should be allowed. Careful management of tourism is needed to prevent problems.

Practice questions

1.

Year	World total tourist arrivals	Earnings from tourism (US$ millions)
1950	25 280 000	2 100
1960	70 000 000	6 900
1970	166 000 000	18 000
1980	286 250 000	105 200
1990	459 200 000	264 700
2000	681 000 000	478 000
2010	1 006 000 000	870 000 (estimated)

a) Describe the increase in tourism since 1950 and compare it with the increase in earnings from tourism.

b) In 2009 the biggest spenders on tourism were from Germany, USA, UK and China. Give possible reasons for this.

2.

Rank	Country	Tourist visitors (million)
1	France	74.2
2	The USA	54.9
3	Spain	52.2
4	China	50.9
5	Italy	43.2

a) The table shows the top 5 countries for tourist arrivals in 2009.
  i) Name one well known attraction in each country.
  ii) Suggest other possible reasons why they have many visitors.

b) Suggest why some countries have greatly reduced numbers of tourists from one year to another. Support your ideas with examples.

3. Complete the following passage, using appropriate words to fill the spaces.

Ecotourism aims for _____ development by _____ the natural environment and _____ the standard of living of local people by providing them with _____ or business.
People visit endangered areas, such as rainforests, in _____ groups, causing as little disturbance and harm as possible to the environment and local people. Where ecotourism is important the trees are _____ cut down for short-term gain as has happened in many tropical forest areas. Ecotourism conserves forest for the _____ because it is an important attraction and economic asset. Ecotourism helps to keep age groups balanced, as the _____ are less likely to move away.

4. Most tourists go on holiday. Write down as many reasons as you can for other tourist travel.

5. List as many different types of holiday as you can.

6. Re-arrange the letters in the anagrams to find terms which match the definitions.

	Anagrams		Definitions
A	roistum	1	A location that attracts many tourists.
B	opts- vero notitendsia	2	A place reached by a journey taking longer than three hours.
C	rueesli ticyivat	3	A place where people on long-haul journeys break their journeys and spend a day or more.
D	horst-ulah aisdnetnoti	4	A person who stays away from home in his or her own country.
E	thoonyep	5	Sustainable tourism which aims to preserve the local environment and cultures while increasing the standard of living of local people.
F	gnol-lahu nositedanit	6	A place reached within three hours.
G	mecrootius	7	Something done for enjoyment in a person's free time.
H	scotidem toirtus	8	Domestic or international travel for any reason for more than a day but less than a year.

7. What has caused the phenomenal growth in tourists? Match the endings to the beginnings to find out.

	Beginnings		Endings
A	Air services…	1	promote their countries' destinations.
B	Cheaper flights…	2	give more leisure time.
C	Reduced airfares…	3	because aircraft are larger and more economical.
D	Shorter working hours and longer holidays…	4	to an increasing number of destinations.
E	Higher wages…	5	allow more to be spent on holidays.
F	Longer life expectancy…	6	has helped people to book cheaper travel deals.
G	The development of package holidays…	7	due to the growth of budget airlines.
H	The Internet…	8	has increased the number of elderly tourists.
I	Increased marketing…	9	companies arranging everything.

8. List as many disadvantages of tourism for local inhabitants as you can think of.

9. Name an LEDC area or country with an important tourist industry.
Name three physical and three artificial reasons why it attracts tourists.

Energy – oil, gas, coal, nuclear and fuelwood

World energy consumption

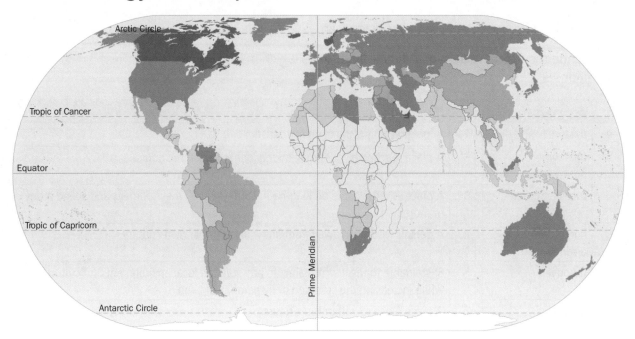

MEDCs have had a bigger share of current world energy consumption. However by 2007 LEDCs were consuming a similar amount. In the future there will be more rapid growth in energy demand expected from the countries that are now LEDCs.

China and India lead the world's economic growth and increase in energy consumption.

The benefits of increased energy consumption

- Electricity makes the daily household tasks easier – it provides heat, light in the evenings, television and computers. People do not have to collect fuel or use candles or lamps at night.
- Modern transport systems are mainly based on oil (petroleum) and allow goods to be moved around the globe and people to travel.
- Industry requires energy to make it work. Without it the economy cannot grow, wealth cannot be increased and people's lives will not be improved.

The problems of increased energy consumption

- Concerns that we are using non-renewable energy sources too quickly and that we will run out of supplies.
- Worries that use of fossil fuels is resulting in air pollution and an increased rate of global warming.
- The inter-dependence of countries on each other for supplies, e.g. of oil and gas, can lead to conflicts.
- Concerns about the safety of nuclear power.

Energy consumption, 2004
kg oil equivalent per person

over 10 000
2500—10 000
1000—2500
250—1000
under 250

no data

Highest energy consumers
kg oil equivalent per person
United Arab Emirates 23 134
Bahrain 21 011
Qatar 15 286
Iceland 13 671
Trinidad and Tobago 12 563

Lowest energy consumers
kg oil equivalent per person
Brurundi 26
Mali 25
Cambodia 15
Afghanistan 14
Chad 8

Fossil fuels

Coal, oil and gas are produced from organic material (plants and animals) which was growing millions of years ago. To grow, this organic matter got its energy from the sun. When we use these fuels we are using the sun's energy from millions of years ago which has been stored in the fossil fuels. It is fossilised energy. Reserves of these **non-renewable** fuels will eventually run out

Oil

Crude oil or petroleum is a mixture of different **hydrocarbons**. Crude oil is found soaked into porous rocks. To extract it, a "well" (which is a **borehole**) is drilled into the ground and the oil will come out either under its own pressure or will need pumping out. **Oil rigs** may be on land or at sea.

The crude oil then requires refining to produce petrol for vehicles ("gasoline" or "gas" in the USA), diesel fuel, aviation fuel and heating oil.

Advantages of oil

- It is easy to transport by pipeline or bulk tanker
- It is the only fuel in mass use for motor vehicles
- It is less polluting than coal when burned
- It is also a raw material in the chemical industry

Disadvantages of oil

- Burning oil produces greenhouse gases and can lead to increased global warming
- Oil spills from leaking tankers and pipelines can cause pollution which kills wildlife
- World oil production is concentrated in a few countries who control the supply and prices
- Work on board oil rigs, especially those offshore, can be dangerous

Natural gas

Advantages of natural gas

- Electricity generation is less expensive with natural gas than with oil, and gas-fired generating plants are less expensive to build than plants that use coal, nuclear, or most renewable energy sources
- The other advantages and disadvantages are the same as for oil

Coal

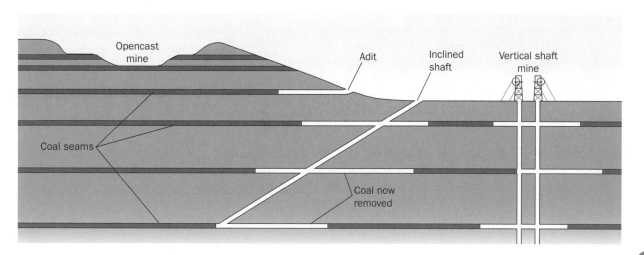

The main problems of deep mining include:

- visual pollution from the coal storage, railway lines and mine buildings at the surface
- the possibility of subsidence where the surface collapses into old workings
- the dangers to the miners from accidents with machinery, gas explosions and roof collapses
- the greater initial capital input compared with open cast mining.

Potential problems from open cast mining include:

- visual pollution from the enormous open pit formed
- the temporary loss of land to other uses during mining
- noise from machinery and blasting
- dust if the pit is allowed to become dry.

Thermal power stations

Factors affecting the location of thermal power stations

- Coal is bulky and is normally transported by rail. For this reason, coal-fired power stations are often located close to the coal mines. Oil and gas can be transported relatively easily by pipeline so the power stations do not need to be located close to the oil wells. Nevertheless oil-importing countries often locate oil-fired power stations at oil refineries close to the port where the oil arrives in the country.
- A supply of cooling water is needed and this is the reason why so many power stations have large cooling towers. Many power stations are located along rivers for cooling water but they must be spaced so that too much warm water is not returned to the river to damage aquatic life. Sea water is not a good coolant because of the salt content which attacks pipework.
- A large flat site is needed for the plant, cooling towers, fuel storage and railway lines.

Advantages of thermal power

- Many countries still have large reserves of fuel, for example coal in South Africa and Germany
- Oil and gas can be transported efficiently by pipeline
- Oil and gas are cleaner than coal in that they produce less air pollution, however they still produce carbon dioxide

Disadvantages of thermal power

- Power stations are major sources of carbon dioxide, nitrogen oxides and sulfur dioxide which are health hazards and contribute to the greenhouse effect. The air pollution caused by one country is often "exported" to another by prevailing winds.
- World coal reserves may only last for another 300 years and oil and gas for an even shorter time.
- Deep mining is dangerous with careful health and safety measures required.
- Over-reliance on imported fuels can cause problems. It makes countries vulnerable to sudden increases in price and political and even military threats from the exporting countries.

Making thermal power stations cleaner

Attempts are being made to develop cleaner coal-fired power stations and cut greenhouse gas emissions. Many developments are being researched but so far few of them have been connected to a full-size power plant.

Nuclear power stations

In a nuclear power station uranium replaces the coal, oil or gas used in thermal power stations. Uranium ore usually occurs in the ground at relatively low concentrations so most is mined by open-pit mining. Only a small number of countries of the world mine uranium. Kazakhstan, Canada and Australia are the top three producers and together account for 63% of world uranium production. Other important uranium producing countries are Namibia, Russia, Niger, Uzbekistan and the USA.

Factors affecting the location of nuclear power stations

- Like other power stations, large flat sites are needed for the plant and for cooling towers.
- The volume of raw material is so small that this is not a factor.
- Pure water for cooling is needed – sea water will not do unless it is desalinated. Sea water has been used in emergencies.
- Some nuclear power stations are built on the coast to dispose of very low level liquid, radioactive waste.
- In countries like the UK, concerns about the safety of early nuclear plants meant that they were located in places far away from areas of dense population, for example Dounreay in the north of Scotland and Calder Hall in Cumbria. This is not the case today.

Advantages of nuclear power

- Only very small amounts of uranium are needed to produce large amounts of energy.
- Uranium ore will not run out in the foreseeable future (some people even classify nuclear power as renewable).
- It does not produce greenhouse gases and acid rain.
- The safety record of nuclear power stations has improved and the industry is highly-regulated.

Disadvantages of nuclear power

- There have been serious incidents at nuclear sites which have lead to leaks of radioactivity, e.g. Chernobyl, Ukraine 1986, Three Mile Island, USA 1979 and Windscale, UK 1957. Radioactivity is a known cause of diseases such as cancer and leukaemia. The earthquake in Japan in March 2011 caused an explosion and leakage of radioactive material at the Fukushima nuclear plant. This raised questions about the safety of nuclear plants in earthquake zones.
- The cost of shutting down old nuclear plants (decommissioning) is very high.
- The radioactive waste from power stations remains a health hazard for hundreds of thousands of years, requires careful storage and is difficult to dispose of safely.

- Nuclear power stations produce material which is the raw material for nuclear weapons.
- The capital costs of building nuclear power stations are extremely high.

Fuelwood in LEDCs

In many LEDCs fuelwood accounts for about 70% of energy supplies. The more rural a country, the greater the dependence on wood. Fuelwood has the advantage that it may be "free" to the user and does not require high technology equipment. It provides an accessible source of fuel for heating and cooking. If there is enough land then wood can be a renewable, sustainable energy source but this is not always the case. It is not a fossil fuel. For rural people who live near towns, surplus wood is often collected and sold to townspeople, sometimes by the roadside, so it becomes a cash crop.

Potential problems with fuelwood use

- In some areas natural woodland is being cut quicker than it can grow back. This means that longer and longer distances have to be walked to collect wood – meaning a lot of hard work and more time being taken up by this.
- Deforestation may lead to exhaustion of soils and soil erosion so that the forest cannot grow back.
- Burning wood in confined spaces on inefficient stoves leads to respiratory illnesses.

In some areas schemes are developed to improve the system of using fuelwood. These usually involve:

- planting more trees, often on a "woodlot" system where there is a constant cycle of re-planting
- managing the woodland and using systems such as careful pruning and thinning to encourage more growth
- the introduction of new fast growing species
- the introduction of new fuel-efficient stoves which cause less smoke.

Renewable energy and water supply

Renewable energy supplies

These include hydro-electricity, geothermal power, wind power, solar power and biofuels. Nuclear power is sometimes classified as renewable.

Hydro-electricity and wind are expected to provide the largest shares of the projected increase in total renewable generation. However, this will vary from country to country.

Geothermal power

Geothermal energy is energy extracted from hot rocks or water beneath the surface. Deeper into the Earth, temperatures rise on average about 25°C for every kilometre. In volcanic areas the increase in temperature may be as much as 70°C for every kilometre. These hot areas tend to be near plate margins and the prospects for geothermal energy are greatest here.

Advantages of geothermal power

- It is extremely cheap and reduces dependence on fossil fuels
- It does not produce greenhouse gases
- The water is pumped back into the ground and re-used
- Unlike other types of renewable energy, it can operate at anytime of the day and year and is not affected by the weather

Disadvantages of geothermal power

- It is restricted to areas with suitable geology
- Areas with suitable geology are sometimes affected by earthquakes and volcanoes
- Although it is usually classified as renewable, each well may only be used for about 25 years
- The ground water is saline and often poisonous

Wind power

Wind farms are located in wide open spaces on land where there is likely to be the strongest winds, e.g. near the coast or on hills. Today offshore wind farms are being built around coasts. Individual businesses may have their own private windmills.

Advantages of wind power

- It does not cause air pollution, global warming or acid rain
- It has very little effect on the local ecosystem, except very occasionally killing birds
- In Europe winds are strongest in winter when there is a peak demand for electricity
- After the initial capital input, production is cheap as the fuel is free
- Wind farms may provide a small source of income for farmers

Disadvantages of wind power

- Wind power cannot be used during calm periods or storms
- Many people consider wind farms to be a form of visual pollution, especially in areas of natural beauty
- The technology is relatively new and at present very large numbers of turbines are needed to generate fairly modest amounts of electricity

Solar power

Light is converted into electricity using **solar panels** (**photovoltaic cells**). When more solar energy is generated than is being used it can be stored in a battery or exported to the national utility grid.

Advantages of solar power

- It is safe and pollution-free
- After the initial capital input, production is cheap as the fuel is free
- It can be used effectively for low power uses such as heating swimming pools or central heating
- Its greatest potential is in warm, sunny countries or in LEDCs where people live in locations which are isolated from the national electricity grids

Disadvantages of solar power

- The initial capital input is high
- It is not as effective in cloudy countries
- It is less effective in high latitude countries where more power is needed in winter but the days are shorter and the sun is lower in the sky giving less light
- It is less effective for high output uses such as powering colour TVs

In many countries the greatest use of solar power has been by private individuals or companies rather than as a contribution to national electricity grids.

Biofuels

These are fuels from biomass. They include liquid fuels (bioethanol and biodiesel), biogas and solid biofuels (including fuelwood, described earlier in LEDCs).

Bioethanol is an alcohol made by fermenting the sugar in plants such as maize. Bioethanol can be used as a fuel for vehicles but it is more frequently added to petrol to increase octane values and improve vehicle emissions.

Biodiesels are made from vegetable oils such as rape seed oil but also from re-cycled used cooking oils from restaurants and kitchens. Like bioethanol, it is often used as an additive to other fuels.

Biogas is methane produced by the breakdown of organic material by bacteria. It can be produced either from waste materials or by the use of energy crops.

Solid biofuels (often referred to as biomass) can be used in power stations and in the heating systems of houses and buildings. Special fuels and boilers are needed.

Advantages of biofuels

- Prices could be more stable than world oil prices
- Supplies can be more secure and reduce reliance on imported fuels
- Fewer pollutants are produced than by than fossil fuels
- It is "carbon-neutral" in that the growing crop absorbs carbon dioxide from the air which balances emissions from the burnt fuel

Disadvantages of biofuels

- In the period from 2008 to 2011, some land previously used for the production of food was changed to produce crops for biofuel production. This led to increases in world food prices and decreases in the food supply.

Hydro-electric power stations

Location

Ideally the site for a hydro-electric power station should have:

- a large river
- a large falling distance (head) of water
- a constant flow of water thoughout the year
- a narrow valley to provide a good dam site
- impermeable rocks so that the reservoir does not leak
- stable geological conditions so that the dam and sides of the reservoir do not collapse and cause a disaster
- sparsely populated land so that large numbers of people do not have to be moved to create the reservoir.

Advantages of hydro-electricity

- Once a dam is constructed, electricity can be produced at a constant rate
- The power stations can respond quickly to changing demand as explained above
- There are no fuel costs
- The lake that forms behind the dam can be used for water sports and leisure activities
- The water can be used for irrigation and other purposes
- There is no atmospheric pollution
- Many HEP projects also supply water

Disadvantages of hydro-electricity

- Dams are extremely expensive to build and they must operate for many decades to make a profit.
- The flooding of large areas of land means that the natural environment is destroyed, along with the natural habitats and historical or archaeological features.
- People living in villages and towns that are in the valley to be flooded must move. In some countries, people are forcibly removed so that hydro-power schemes can go ahead.
- The building of large dams can cause serious geological damage.
- Although modern planning and design of dams is good, in the past old dams have been known to collapse. This has led to deaths and flooding.

- When a river flows from one country to another a dam in one country affects the flow of the same river in the next country and can lead to serious disputes between neighbouring countries.
- Dams catch sediment that would have flowed down the river and increased the fertility of soils when it settled from flood waters. The sediment also reduces the capacity of the dams.

Water supply

Water supply comes from two different types of source, both of which come from rainfall (or snow melt).

Surface water is water from rivers and lakes. Because river flow is sometimes variable, rivers are often dammed to create reservoirs which store water for dry periods.

Groundwater is water held within the spaces in porous, permeable rocks, in the same way that oil and gas is held. Layers of rock which contain water are called **aquifers**. The water is extracted either by digging wells or drilling boreholes. An electric pump brings the water to the surface. This may use wind or solar power.

Water is in demand for:

- **Agriculture** – irrigation in dry areas.
- **Domestic use** – In MEDCs people use large volumes of water each day for washing, flushing toilets, watering gardens and even washing cars. In many LEDCs this luxury is not available.
- **Industrial use** – Many industries use large volumes of water in processing (e.g. paper manufacture) and cooling (e.g. power stations).

The balance between these uses varies greatly from area to area.

Water deficit is where water demand exceeds supply.
Water surplus is where supply exceeds demand.

Factors affecting whether there will be a surplus or a deficit of water

- Amount of precipitation.
- Temperatures and amount of evaporation. The higher the temperatures, the greater the amount of rainfall which evaporates before it can be used.
- Importance of irrigation. Where irrigation is practised over large areas it is common for agriculture to be the biggest user of water compared with industrial or domestic use.
- Level of economic development. In industrialised countries, large amounts of water are used by factories or for cooling water in thermal or nuclear power stations.
- Population density. Large conurbations use large amounts of water for domestic use.
- Presence of water bearing rocks. Even in deserts, the presence of aquifers below the surface, fed by rainfall outside the desert, can provide adequate water supplies.
- Proximity to rivers. Some areas use rivers fed by rainfall thousands of kilometres away.

The need for sustainable development, resource conservation and management

Agriculture, mining, quarrying, energy production, manufacturing industries, transport and tourism improve the quality of our lives by providing employment and generating wealth for both the worker and the country.

Many resources are finite – once used up or destroyed beyond repair, they are gone forever. Careful management to conserve them for future generations is essential. There are concerns that we are using non-renewable energy sources too quickly and that we will run out of supplies. The inter-dependence of countries on each other for supplies, e.g. of oil and gas, can lead to conflicts. World coal reserves may only last for another 300 years and oil and gas for an even shorter time.

In some areas natural woodland is being cut quicker than it can grow back. Deforestation may lead to exhaustion of soils and soil erosion so that the forest cannot grow back.

Soil is a very important resource, as it allows crops and vegetation, necessary for human and animal survival, to grow.

Soil erosion

Attempts to increase agricultural production and food supply must be sustainable - done in a way that can be continued into the future. Soil takes a very long time to form but can be very rapidly removed by erosion.

Soil erosion occurs when:

* the soil is exposed and not covered by vegetation or crops
* the soil, damaged by poor agricultural practices, is loose and has lost its structure.

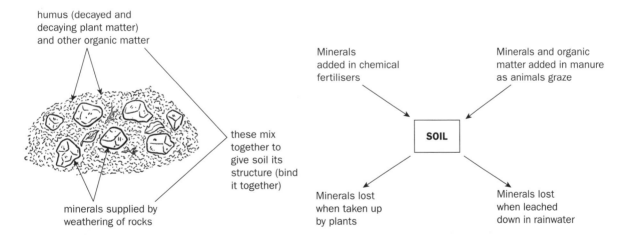

Soil erosion by wind

In areas with low rainfall the soil dries out and becomes loose. Strong winds can then remove it. Erosion can be caused by poor agricultural practices, such as ploughing where rainfall is unreliable, and monoculture which destroys the soil structure by removing the nutrients. The soil dries out, turns to dust and is easily blown away.

Soil erosion by running water

This occurs on steep slopes when rainfall is so heavy that all of it does not soak into the ground and surface runoff occurs down the slopes, either in sheets of water (on gentler slopes) or concentrated into channels forming gullies (on steeper slopes). It is often started by poor farming methods.

Soil conservation methods

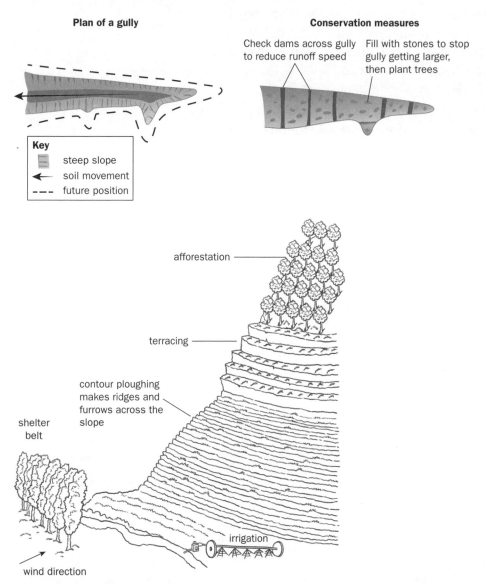

Plan of a gully

Conservation measures

Check dams across gully to reduce runoff speed

Fill with stones to stop gully getting larger, then plant trees

Key
- steep slope
- ← soil movement
- - - - future position

afforestation

terracing

contour ploughing makes ridges and furrows across the slope

shelter belt

irrigation

wind direction

▲ Some soil conservation methods

Conservation method	Erosion prevented		Description
	Wind	**Water**	
Crop rotation	✓	✓	A different crop is grown on a plot each year for three or four years before the first crop is grown again. As different crops take different nutrients from the soil the soil does not become exhausted, lose its structure and become loose and easily eroded.
Fallow periods	✓	✓	A piece of land is "rested" every few years to allow it to regain lost nutrients. The soil therefore keeps its structure. (Only a wise measure if in an area where a vegetation cover will grow quickly.)

Strip cultivation and inter-cropping	✓	✓	Crops are grown in narrow bands, often at right angles to the prevailing wind, with other bands of different crops in between. The crops are harvested at different times so the field is never completely bare. Soil, blown by the wind from a bare strip, is trapped by the crop in the next strip.
Cover cropping	✓	✓	This works on the same principle as strip cultivation. Usually a fast growing crop is planted after the main crop has been harvested. Sometimes this is a "green manure" crop which is ploughed back into the soil to add nutrients. In this way the soil is left bare for the minimum time.
Reducing stock density	✓	✓	By having fewer livestock, a piece of land does not become overgrazed. There is always a cover of vegetation to protect the soil. Fencing fields to make paddocks allows rotational grazing and stock density to be kept to carrying capacity.
Dry farming (This is practised in the drier parts of the Canadian Prairies where wheat is grown.)	✓		These methods aim to reduce water loss from the soil by not ploughing the land, covering the soil with a mulch, growing drought resistant varieties, planting crops in alternate years and sowing seeds into the previous year's stubble.

Practice questions

1. State one reason why each of the following helps to increase quality of life by creating wealth: mineral extraction, energy production, manufacturing industry, transport, tourism.

2. List ways in which the following resources can be conserved:
   a) Fossil fuels
   b) Mineral ores
   c) Forests.

3. Sort this list into pairs of opposites.

   crop rotation, deforestation, ploughing up and down a slope, irrigation, reforestation, contour ploughing, dry farming, monoculture, cover cropping, fallow.

4. Fill in the blank spaces about soil erosion in subsistence farming areas in Swaziland. Use the words in the box below:

   > Bare, down, gullies, loose, monoculture, over-cultivation, overgrazing, structure, terraced, up

   The steeper slopes in the High Veld are _____
   where crops are grown but on some steep slopes _____
   have been formed, as it rains very heavily in the wet season and the
   sandy soils are naturally _____ In places poor farming
   practices have added to the problem:

(continued on next page)

- The soil _____ has been destroyed by _____ and by growing maize in a _____ .
- Ploughing _____ and _____ slopes channels runoff, taking soil with it.
- Pasture burning leaves the soil _____ .
- This also results from overstocking of the pasturelands which causes _____ .

5. Make a table with the same headings as on page 138 and use the diagram of these methods to complete it for terracing, contour ploughing, afforestation, shelter belts (wind breaks) and irrigation.

6. Match the terms with their definitions.

	Term		Definition
A	Rotational grazing	1	The removal of minerals from the soil in solution.
B	Soil erosion	2	A small wall built across a gully to stop the force of water running down it and causing further erosion.
C	Organic matter (in soil)	3	An organic substance consisting of partially or wholly decayed vegetable or animal matter that provides nutrients for plants, increases the ability of soil to retain water and gives it good structure.
D	Soil exhaustion	4	When minerals in the soil have become depleted due to over-cropping, resulting in decreasing crop yields.
E	Leaching	5	Method of preventing soil erosion. Crops are grown in narrow bands, with other bands of different crops in between. The crops are harvested at different times so the field is never completely bare.
F	Manure	6	The way in which the particles in the soil group together. A good one is when the particles are not so loose as to be easily eroded but spaces allow water and air movement, worms and micro-organisms to be active, root growth and seedling emergence.
G	Check dam	7	Having a series of fenced areas and moving animals from one area to another to prevent overgrazing.
H	Humus	8	The removal of the soil by wind or running water on slopes.
I	Strip cultivation	9	Producing goods in a manner that is capable of continuing well into the future.
J	Soil structure	10	Any organic matter added to the soil to improve its structure and mineral content, e.g. animal dung or vegetable waste.
K	Sustainable production	11	Chemical elements, some of which are critical to plant growth, e.g. nitrogen, potassium, phosphorus.
L	Fertiliser	12	Chemicals added to the soil to make it more productive. It is often used to mean artificial, industrially produced chemicals.
M	Shelter belt	13	Material from animals, e.g. dung and plants such as decaying roots and leaves.
N	Minerals (in soil)	14	Method of preventing soil erosion. Rows of trees grown on the side of the field at right angles to the prevailing wind. The trees reduce the speed of the wind so that it is not strong enough to pick up the soil.

The impact of pollution on the natural environment

GLOBAL WARMING

Greenhouse gases e.g. carbon dioxide (from burning fossil fuels and wood), and methane keep heat in and reflect it back to Earth. Greenhouse gases from industry may be causing enhanced global warming.

This could cause:
- greater climatic extremes (high winds and heavy storms)
- melting of ice taps, causing seas to rise and flood low-lying land
- greater evaporation from soils, causing crop losses
- bleaching of coral reefs
- changing boundaries of vegetation and crop growth

Areas likely to be affected most:
- polar regions
- coasts
- desert fringes

DAMAGE CAUSED BY EXCESSIVE USE OF PESTICIDES

Some pesticides cause damage when they enter the food chain because the chemicals accumulate in the bodies of the higher predators.

Areas likely to be affected most:
- downwind of and downstream of farming areas

POLLUTION FROM SEWAGE AND FERTILIZERS

Sewage and fertilizers from farmland enter lakes and rivers. The nutrients (nitrates and phosphates) make the algae in the water grow rapidly. Other plants are smothered and die.

Bacteria use up oxygen from the water and so fish and other animals die.

This process is called **eutrophication**.

Areas likely to be affected most:
- downstream of urban and intensive arable farming areas

NUCLEAR FALL OUT

Radiation leakage after accidents at nuclear power stations causes long lasting air pollution. The radioactive waste remains a health hazard for hundreds of thousands of years and is difficult to dispose of safely.

Areas likely to be affected most:
- immediately around the damaged plant and on the side to which the prevailing wind blows

ACID RAIN

Combustion of fossil fuels in thermal power stations, industry, homes, and car engines produces sulfur dioxide and gaseous oxides of nitrogen. When these dissolve in water, acids are formed. These fall as acid rain.

Acid rain makes the soil acidic. This damages plant roots and encourages the leaching of essential calcium and potassium minerals from the soil and their replacement with aluminium and manganese which are harmful to roots. Plants and trees may be killed. Acid rain and chemicals washed from the soil enter rivers and lakes. This kills fish and other aquatic animals.

Areas likely to be affected most:
- forests
- lakes and rivers in areas to which the prevailing wind blows from large industrial and urban areas

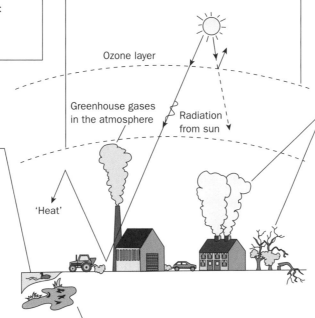

Ozone layer

Greenhouse gases in the atmosphere

Radiation from sun

'Heat'

In MEDCs strict laws ensure that dangerous waste does not normally enter the air, seas or rivers or contaminate the land.

Air pollution

In MEDCs, strict regulation of vehicles and industrial plants has greatly reduced air pollution. Methods of reducing traffic congestion help too. Many modern industries use scrubbers on chimneys to remove harmful gases.

It is in the major cities of LEDCs and NICs that the highest levels of air pollution occur, especially in countries where industrialisation is rapid, such as China and India. New fuel-efficient stoves which cause less smoke are helping reduce the pollution that resulted from cooking using fuelwood.

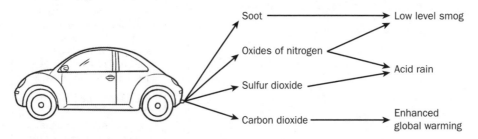

Water pollution

This is a problem affecting aquatic life and water supplies.

- In areas where there is no sewage system, such as some shanty settlements, raw sewage is simply dumped or left in open drains. In other areas the raw sewage is not treated and empties from sewers into rivers or the sea. The solution is for proper sewage pipes to be installed and for the sewage to be treated to make it safe before it is released into rivers or seas.
- Some liquid industrial waste is discharged into rivers from factories. Fifty years ago many of the urban rivers of Europe were "dead". Strict regulation of industry has improved this and fish have returned to many rivers.
- In some LEDCs solid domestic waste is dumped in rivers.
- Even high-tech industries can pollute. Accidental spills and leaks of solvents and acids can cause toxic substances to pollute both the air and water.
- Oil spills from tankers and pipelines can cause river and sea pollution which kills wildlife. The marine ecosystem is being damaged by shipping, especially ships carrying oil and oil products in and out of ports. A lot of oily discharges are pumped out within port areas in LEDCs.

Methods of cleaning up marine oil spills

Booms
Floating inflatable tubes prevent slicks from spreading.

Detergent sprays
Chemicals break up oil into droplets, dispersing larger slicks.

Skimmer
Oil drawn up absorbent belt. Rollers scrape and squeeze oil into collecting tank.

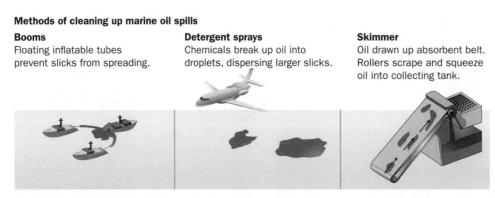

Visual pollution

There are many examples of this. Power stations and factories are ugly. Deep mining needs mineral storage, mine buildings and waste heaps at the surface. Opencast mining has enormous open pits and heaps of soil and rock being stored.

Noise pollution

During quarrying and opencast mining, noise pollution results from blasting and the use of large-scale machinery.

Sources of noise pollution in urban areas include vehicles, trains, aircraft taking off and landing, factories, large congregations of people and noise in residential areas, for example from radios and parties. The problems are worst where cities have grown rapidly without proper planning.

In some cases noise can be seriously disturbing for wildlife. The problem has been addressed in some countries by laws which limit noise from factories and homes.

Practice questions

When answering questions on this topic, always remember to specify the type of pollution you are writing about by inserting air, water, noise or visual before the word 'pollution'.

1. a) Which types of pollution are international problems that require international solutions? Explain your answer.

    b) Why is it more difficult to reduce pollution from individual factories in LEDCs than in MEDCs?

2. The table gives information about some greenhouse gases (n/a indicates 'not applicable').

Greenhouse gas	Amount in atmosphere 2010 (ppm)	Increase since 1750 (ppm/year)	Approximate lifetime of the gas in the atmosphere (years)	Contribution to the Greenhouse Effect (Approximate %)	Global warming potential over a 20 (and 100) year period
Carbon dioxide	365	113	160	17.5	1(1)
Methane	1.7	1.0	11	6	72 (25)
Nitrous oxide	0.3	Very small	120	Less than 5	289 (298)
Water Vapour	10,500 (average)	n/a	n/a	54	n/a

a) Which gas contributes the most to the natural greenhouse effect?

b) Which two gases have the greatest global warming potential over 20 years, and over 100 years?

c) Why is there more concern about the addition of carbon dioxide to the atmosphere than about the gases you named in (b)?

d) State the difference between the greenhouse effect and the enhanced greenhouse effect.

e) How does farming contribute to the methane?

3.

C	A	R	B	O	N	D	I	O	X	I	D	E	J	A	A	Z	B	O	M	Y
O	B	H	A	R	T	T	Y	A	M	N	Y	K	B	O	D	G	N	H	E	A
V	I	N	D	F	A	L	L	O	W	G	U	M	R	A	H	G	Y	Y	T	E
E	S	A	R	F	A	R	O	S	H	G	U	L	L	Y	N	E	E	D	H	M
R	E	W	I	N	D	A	E	G	E	B	N	Y	H	J	U	O	E	R	A	A
C	C	H	H	O	R	N	Y	B	U	E	D	S	L	O	M	T	A	O	N	B
R	G	A	S	T	C	O	R	S	E	U	I	N	D	E	L	H	I	E	E	Y
O	M	N	U	C	L	E	A	R	S	T	O	M	P	U	R	E	I	L	E	B
P	Y	A	M	U	N	A	P	A	R	R	B	A	N	G	A	R	E	E	G	I
P	A	M	U	B	O	O	M	S	A	O	C	A	I	R	O	M	E	C	T	P
I	M	U	R	A	D	A	B	A	D	P	U	T	T	A	R	A	B	T	N	S
N	C	H	S	K	O	N	I	C	H	H	A	M	L	I	H	L	E	R	B	U
G	M	A	T	N	Z	E	N	E	U	I	M	U	N	N	I	J	E	I	N	S
S	W	J	R	W	O	M	I	E	O	C	H	Y	D	R	A	B	A	C	D	T
H	I	A	U	N	N	B	T	L	E	A	C	H	I	N	G	J	T	I	A	A
E	N	P	C	K	E	H	I	N	D	T	A	N	R	H	A	B	C	T	B	I
E	D	A	T	Y	C	R	O	O	S	I	N	T	H	E	J	U	N	Y	N	N
L	S	N	U	K	L	O	B	L	Y	O	W	O	F	D	I	N	U	S	A	A
A	E	D	R	Y	F	A	R	M	I	N	G	C	H	A	M	P	U	K	A	B
G	D	N	E	M	U	L	S	C	R	U	B	B	E	R	S	J	D	E	Y	L
S	E	A	L	E	V	E	L	N	A	S	R	U	D	D	I	N	K	H	A	E

Find the terms in the wordsearch which match the following descriptions about environmental risks and benefits.

a) The main gas responsible for enhanced global warming.

b) Ruminant animals are one source of this powerful greenhouse gas.

c) This term is used to describe the problem of excessive nutrient content in rivers.

d) This is produced when soil erosion is channelled.

e) The term used for resting a piece of land.

f) This method of reducing soil erosion involves planting another crop between the rows of the main crop.

g) This term includes a number of ways in which the farmer aims to reduce water loss from the soil.

h) This term means how the soil particles are held together and has a different meaning when used in connection with forests and with rocks.

i) This type of power station produces dangerous and long lasting waste.

j) These are responsible for acid rain falling far from the source of the pollutants.

k) This degrades soils and is encouraged by acid rain.

l) This gas produces air pollution when at ground level but is a valuable gas in a layer high in the atmosphere.

m) This type of renewable energy requires great expense to set up.

n) This type of energy production cannot operate at both extremes of the weather it needs.

o) These are put in factory chimneys to clean pollutants from the emissions.

p) These are placed in water to stop the spread of oil spills.

q) This type of energy production is found in volcanic areas.

r) Global warming is likely to cause a rise in this.

s) This describes what human activities need to be to ensure their children's future.

Theme 3: Economic development and the use of resources

Exam-style questions

1. **a)** Study the diagram which gives information about industry in part of Brazil.

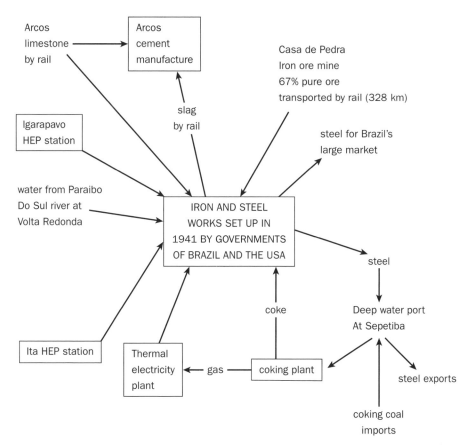

Arcos limestone by rail → Arcos cement manufacture

Casa de Pedra Iron ore mine 67% pure ore transported by rail (328 km)

slag by rail

Igarapavo HEP station

steel for Brazil's large market

water from Paraibo Do Sul river at Volta Redonda

IRON AND STEEL WORKS SET UP IN 1941 BY GOVERNMENTS OF BRAZIL AND THE USA

steel

coke

Deep water port At Sepetiba

Ita HEP station

Thermal electricity plant ← gas ← coking plant

steel exports

coking coal imports

   i) Explain how the building of the steelworks in 1941 in Brazil, an LEDC, was financed. [1]

   ii) Name the input to the iron and steel works that is also a raw material for the cement industry. [1]

   iii) Name two outputs from the iron and steel works. [1]

   iv) Explain why waste is not named as an output from the industries shown on the diagram. [2]

   v) Suggest why the iron and steel works has low energy costs. [2]

   vi) Suggest **other** reasons why production costs at the iron and steel works are low. [3]

   b) Explain why transport costs and nearness to raw materials have become less important today as factors influencing industrial location. [3]

   c) With reference to examples, explain why some industries pollute the environment. [5]

   d) For a country you have studied where high technology industry is important, describe the factors that have influenced the location of the high technology industry there. [7]

2. a) Study the information about desertification in China.

> There is more desertified land in China than in any other country. By 1990
> It affected 30% of the land, especially in semi-arid areas across the north,
> where in the 1970s and 1980s temperatures increased and precipitation
> decreased. There has been both wind and water erosion. China has 22%
> of the world's population but only 7% of its arable land. Soil degradation
> has occurred, especially near irrigated oases and in farmed sandy areas. The
> increasing population's demand for water has also been a factor.
>
> After the world's biggest tree planting exercise, together with the compulsory
> movement of millions of nomadic herders from affected areas and restrictions
> on farming, desertification has been stabilised and some improvements
> made. The deserts are shrinking - but at the present rates it will take
> 300 years to restore the 530 000 square kilometres that can be treated.

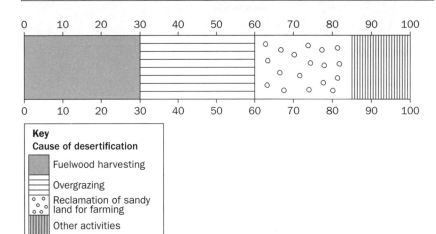

Key
Cause of desertification

- Fuelwood harvesting
- Overgrazing
- Reclamation of sandy land for farming
- Other activities

  i) According to the divided bar graph, what percentage of
  desertified land was caused by agricultural activities? [1]

  ii) Explain why desertification particularly affected semi-arid areas
  with sandy soils between 1970 and 1990. [3]

  iii) Explain the fact that 1.6 billion tonnes of soil flow down China's
  Yellow River each year. [3]

  iv) Explain why many of China's cities have been badly affected by
  sandstorms from time to time. [3]

  v) Find in the information two reasons why the increase of an
  already large population resulted in desertification. Explain how
  each of them would cause desertification. [4]

b) Soil degradation occurs as a result of the soil losing its structure.
Explain what this means and how it occurs. [4]

c) Choose an area you have studied where measures have been taken
to conserve the soil. Name the area, describe the measures taken
and explain why they should be effective. [7]

3. a) **i)** Explain the difference between *pastoral* and *arable* farming.   [1]
     ii) Explain the difference between *intensive* and *extensive* agriculture.   [2]
     iii) List three *processes* of an agricultural system.   [3]
     iv) Describe the differences in the inputs of a *commercial* farm and
     a *subsistence* farm.   [4]

b)

Inputs	Borneo	Japan	California
Direct energy			
Labour	0.626	0.804	0.008
Axe and hoe	0.016		
Machinery		0.189	0.360
Vehicle fuel		0.910	3.921
Indirect energy			
Fertilizers		2.313	4.317
Seeds	0.392	0.813	1.140
Irrigation		0.910	1.299
Pesticides		1.047	1.490
Electricity		0.007	0.380
Transport		0.051	0.121
Total inputs			
Output-rice yield	**7.318**	**17.598**	**22.3698**
Energy efficiency ratio	**7.08**	**2.49**	**1.57**

(Source: Byrne, K 2000. *Environmental Science, Bath Science.* Cheltenham UK. Nelson Thornes.)

▲   Energy efficiency ratios for rice production (Thousands of kilocalories per hectare)

    i) Using the table, describe the differences in rice yield (output) per
    hectare in Borneo, Japan and California.   [3]
    ii) Using the table suggest reasons for the differences in rice yield
    between Japan and California.   [5]

c) For a named agricultural system that you have studied, describe the
physical and human inputs.   [7]

4. a) **i)** What is meant by the term *fossil fuel*?   [1]
     ii) Describe two disadvantages of fossil fuels.   [2]
     iii) Name three renewable fuels.   [3]
     iv) Describe the disadvantages of using some renewable fuels.   [4]

b)

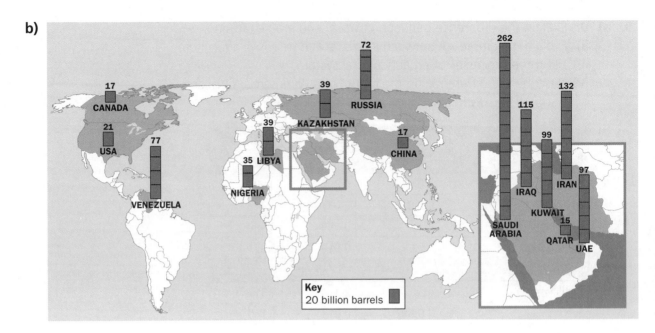

i) Look at the map of major world oil reserves. Which continents have large reserves and which have small reserves? [3]

ii) Compare the world distribution of major oil reserves with the distribution of demand for oil. How is oil transported between these areas? [5]

c) For an example of a hydro-electric power scheme that you have studied, explain the advantages of the area for a hydro-electric power station and any problems that the scheme has caused. [7]

Survey maps (topographic maps)

These are examples of large-scale maps. This means that they show a relatively small area of land in great detail. They show the surface features of an area including relief, drainage, land use, settlement and roads. This is a compulsory element of the CIE IGCSE Paper 2.

Using the key and symbols

The positions of different features on a map are shown by symbols. Different countries use different symbols on their maps, so it is always best to check the meaning of a symbol using the key which is a list of the meaning of each symbol, usually at the side or at the bottom of the map.

Map scale

The scale of a map shows how distance on the ground has been represented on the map. A large-scale map might show a small area such as a school or a village, whereas a small-scale map might show a whole country. Scale is shown by the representative fraction and the scale line.

Representative fraction

		Distance on map	Distance on ground
Representative fraction	1 : 25 000	1cm	25 000cm
		4cm	1km
	1 : 50 000	1cm	50 000cm
		2cm	1km

Scale line

SCALE 1:25,000

Distance measurement

Remember the method of measurement using the edge of a sheet of paper and the scale line. Avoid calculations!

Eastings and northings

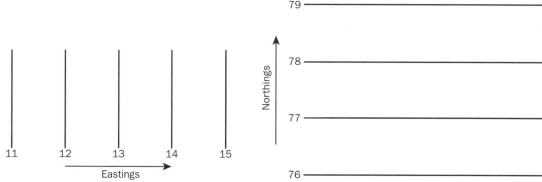

Four figure grid references

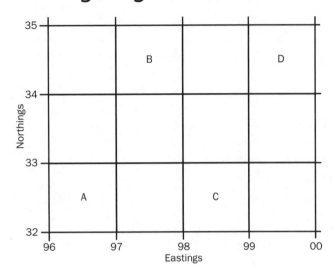

The four figure grid reference of A is 9632, B 9734, C 9832, D 9934.

Six figure grid references

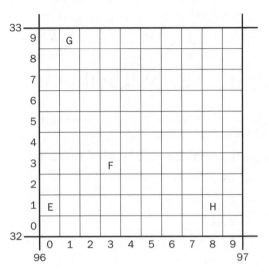

The six figure grid reference of E is 960321, F is 963323, G is 961329, H is 968321.

Compass directions

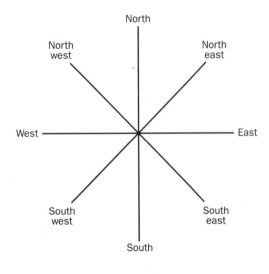

360° bearings

This is an accurate way of showing direction clockwise from grid north. Use a protractor exactly over the point you wish to measure from and aligned north – south along the grid lines as shown below.

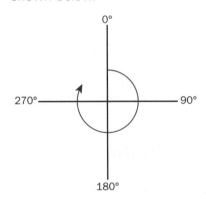

Spot heights and trigonometrical points

A dot on the map with a number beside it shows the number of metres that the point indicated by the dot is above sea level. Sometimes the spot height is combined with a trigonometrical point (station).

Contours

A contour on a map is a line, often brown in colour, which joins places of equal height above sea level. The difference in height between the contours (sometimes called the contour interval) varies but is often 10 metres or 20 metres. Important contours such 100 metres, 200 metres, 300 metres etc. are often shown by a bold line.

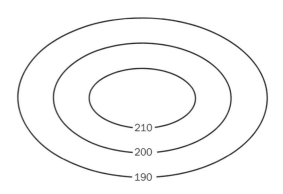

Gradient

Measure the horizontal distance between two points. Find the difference in height between the two points from spot heights or contours.

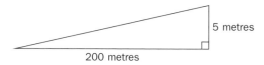

5 metres

200 metres

$$\text{Gradient} = \frac{\text{Vertical Interval (difference in height)}}{\text{Horizontal Equivalent (horizontal distance)}}$$

$$= \frac{5 \text{ metres}}{200 \text{ metres}}$$

$$= \frac{1}{40}$$

$$= 1 \text{ in } 40 \text{ or } 1 : 40 \text{ or } 2.5\%$$

Practice questions

1. **a)** How big are the grid squares on a 1:25 000 map?
    b) How big are the grid squares on a 1:50 000 map?
    c) Is an atlas map a large-scale map or a small-scale map?
    d) Is a survey map a large-scale map or a small-scale map?

2.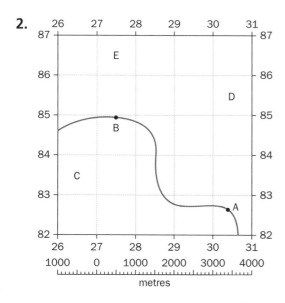

A road is shown crossing the map. What is the distance in metres between points A and B:

a) in a straight line?

b) along the road?

3. Give the four figure grid references for squares C, D and E on the same map.

4.

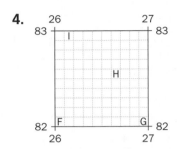

Give the six figure grid references for points F, G, H and I.

5. Look at the map for question 2.

   a) What is the compass direction from point A to point B?
   b) What is the compass direction from point B to point A?
   c) What is the 360° bearing from point A to point B?
   d) What is the 360° bearing from point B to point A?

6. Point A is at 370 metres above sea level and point B is 420 metres above sea level. Using this information and your answers to question 2, calculate the gradient between the two points:

   a) in a straight line.
   b) along the road.
   c) Which is the steeper of the two gradients?

7. Match the following terms with their definitions.

	Term		Definition
A	Scale	1	How distance on the ground has been represented on a map.
B	Gradient	2	A map which shows a relatively large area of land in less detail, for example a map of the world.
C	Eastings	3	A list of the symbols shown on a map which shows what they mean.
D	Representative fraction	4	A map which shows a relatively small area of land in great detail, for example a topographic survey map.
E	Trigonometrical point	5	Lines on a map which join points of equal height above sea level.
F	Grid reference	6	Lines running north – south on a survey map.
G	Spot heights	7	A precise measure of the steepness of the slope.
H	Northings	8	A system which allows locations on a map to be described precisely by 4 or 6 figures.
I	Key	9	This is a pillar about a metre tall which is used as a fixed point by the mapmakers. It is shown on a survey map along with its height above sea level.
J	Large-scale map	10	A dot on the map with a number beside it to show the number of metres that the point indicated by the dot is above sea level.
K	Contours	11	Lines running east – west on a survey map.
L	Small-scale map	12	A way of describing the scale of a map. It shows how many units of distance on the ground are shown by one unit of distance on the map.

Physical features on survey maps: relief and drainage

Human features on survey maps: settlement, communications and landuse

Relief features

The geographical term relief means the height, steepness and shape of the ground surface.

Slopes

The closeness of the contours shows the steepness of the slope. Closely spaced contours means a steep slope, widely spaced contours means a gentle slope. The absence of contours may indicate flat land. Cliffs are shown by a separate symbol.

Uplands and lowlands

The contour heights and spot heights on the map show the height above sea level and can be used to show the higher and lower areas on a map. There is no precise definition as to how high or low an area has to be to be classified as highland or lowland. In some areas of the world entire countries are high above sea level.

Valleys and flood plains

Small valleys without a flat floor are shown on maps by a "V" shape in the contours. The V always points to high ground. There may or may not be a river in the centre.

Plateaux

This is land which is high and flat.

Ridges

A ridge is a long, narrow area of high ground, rather like the spine of an open, upturned book.

Spurs

A spur is a ridge where the spine slopes down from high ground to low ground. A spur is shown by a "V" shape in the contours, where the V points to low ground.

Scarps

A scarp is broad, steep slope. It could be the sides of a plateau or a ridge. The slopes may include cliffs.

Describing relief on maps

- Is the area highland or lowland?
- What is the average height and the height of the highest point?
- Are there areas which are steep, gentle, flat or cliffs?
- Are there any features such as valleys, flood plains, plateaux, ridges, spurs or scarps?

Drainage

Drainage means the water features shown on the map, usually in blue. This includes rivers and streams and their features, lakes and ponds. Marsh may be considered to be a feature of the drainage, i.e. poorly drained land, or it may also be considered a feature of the vegetation where the plants are adapted to these conditions. Drainage may also include features produced by human activity such as drainage channels, dams or reservoirs.

Drainage density

Drainage density is the total length of the rivers and streams in an area in kilometres divided by the area in square kilometres. Areas of high drainage density have lots of surface water and areas of low drainage density have very little, often due to permeable rocks such as limestone which cause the water to seep underground.

Channel shape

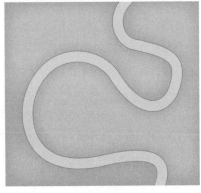

Meandering

Straight

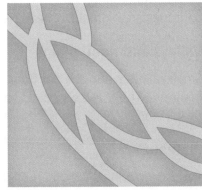

Braided

Stream (drainage) pattern

Dendritic

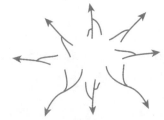

Radial

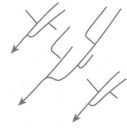

Trellised

Describing drainage on maps

- Is there a main river, what is its width and are there small streams?
- What is the flow direction?
- Is the drainage density high or low?
- What is the channel shape?
- What is the drainage pattern?
- Are there features such as dams, waterfalls, rapids?

Cross sections

A cross section is a type of diagram often used in examination questions. It is as if a giant has sliced the landscape vertically along a line and pulled it apart. Cross sections are drawn to scale. The horizontal scale is generally the same as the map scale but the vertical scale is made bigger (vertical exaggeration) so that features such as hills and valleys show up better. The position of features on the ground surface can be shown with labelled arrows.

Physical features of coastlines

Are there features such as:

- bays
- headlands
- estuaries or river mouths
- beaches
- coral reefs
- cliffs
- wave-cut platforms
- stacks
- marsh or swamp

Urban morphology

Urban morphology refers to the form of towns and cities, or the variations in land use within them. Different urban zones can be identified on survey maps.

Nucleated, dispersed and linear settlement

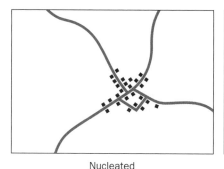

Nucleated

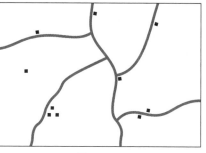

Dispersed

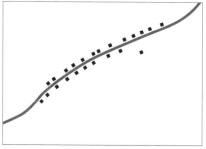

Linear

Distribution, density and location of settlements

These are usually affected by factors such as:
- availability of communications
- accessibility of points such as road junctions (route centres) and bridge points
- availability of cultivated land
- avoiding steep slopes
- avoiding land which is liable to flooding and may also be affected by pests and disease.

Communications

Communications on maps are generally the different types of roads, tracks and railways and, occasionally, ports and airports/airstrips. Care should always be taken to read the map key carefully to identify these features correctly.

Links with physical and human features

Roads usually try to follow gentle slopes. They try to avoid steep slopes and areas which are liable to flood. For this reason, they often follow valleys, at the bottom of the valley sides and avoiding the flood plain. When steep slopes are encountered roads may zig zag and have hairpin bends to make the gradient more gentle.

Railways need very gentle gradients. They often have cuttings or tunnels through hills or cross lowland areas on embankments. Make sure that you know the difference between a cutting and an embankment and the symbols used for them.

Land use

The land use symbols used on maps vary greatly from country to country. Typically they show natural vegetation, types of cultivation and settlement.

Practice questions

1.

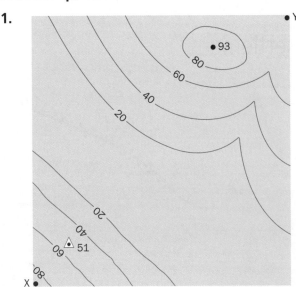

Describe the relief of the area shown on the map.

2.

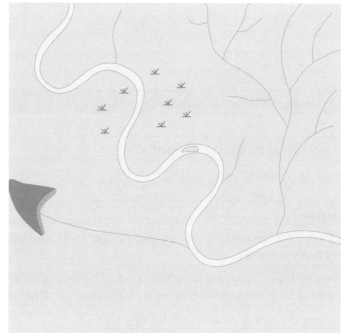

Describe the drainage of the area shown on the map.

3.

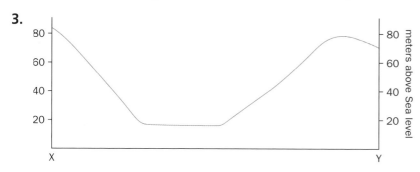

The cross section has been drawn from X to Y across the map used for question 1. On the cross section, use labelled arrows to show the positions of: the flood plain, and a trigonometrical station at 51 metres.

4.

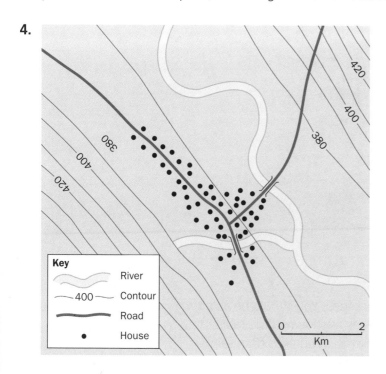

Key

～～	River
—400—	Contour
▬▬	Road
●	House

Look at the settlement shown on the map. How has the location of the settlement been affected by
a) relief
b) drainage and water supply
c) transport?

Photographs, field sketches, data tables and graphs

Photographs

Term	What to describe in your answer to photograph questions
Physical features	Relief, drainage, vegetation
Human features	Features of buildings and settlement, agriculture, industry, transport
Relief	Features of the height and shape of the ground surface, including the names of any features you can identify
Drainage	Rivers, streams, lakes, and their features
Agriculture	Animals, the plots of land, fences, what is in the plots, e.g. grass, ploughed land, bare land, crops, any farm buildings and machinery that you can see
Settlement	Features of the buildings themselves (as listed for housing below), the types of buildings, the use of the buildings and the spacing of the buildings and whether they are nucleated, linear or dispersed
Housing	Size, number of storeys, building materials, quality, windows, the building plots

Field sketches

In examination questions, field sketches are often used with photographs. Remember to labels your sketches as follows.

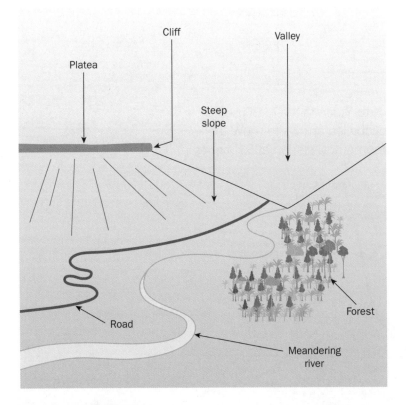

Data tables

Tables of data are often used in examination questions. You should be able to look at the data and identify any patterns or trends as shown in the table on the following page. This shows the number of births and deaths (thousands) in the UK. From 2011, the figures are projected rather than actual.

Year	1951	1961	1971	1981	1991	2001	2011	2021	2031	2041	2051
Births	790	940	900	740	790	670	780	780	770	830	840
Deaths	610	630	650	660	640	600	560	560	630	720	760

Graph types

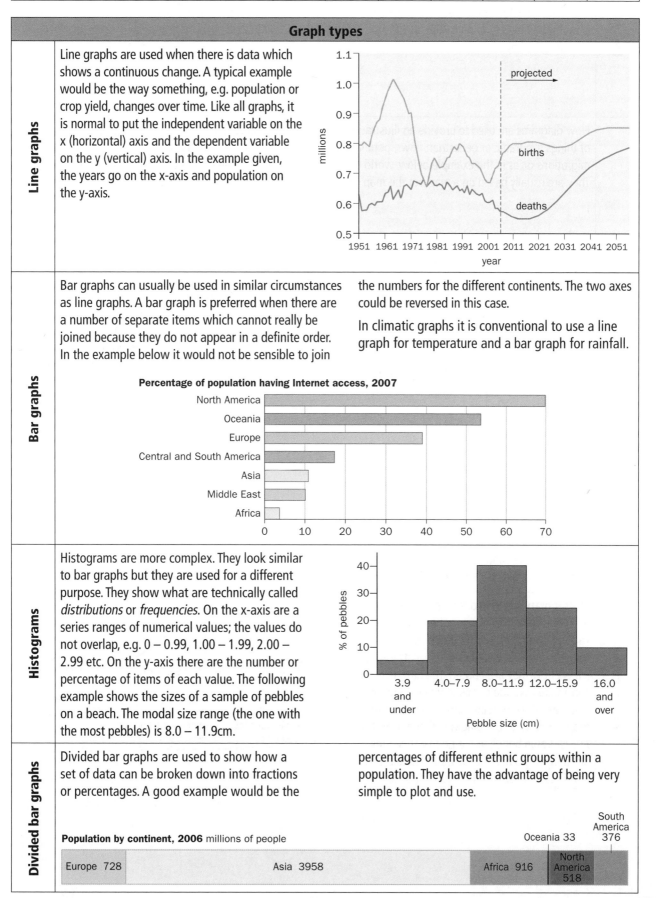

Line graphs

Line graphs are used when there is data which shows a continuous change. A typical example would be the way something, e.g. population or crop yield, changes over time. Like all graphs, it is normal to put the independent variable on the x (horizontal) axis and the dependent variable on the y (vertical) axis. In the example given, the years go on the x-axis and population on the y-axis.

Bar graphs

Bar graphs can usually be used in similar circumstances as line graphs. A bar graph is preferred when there are a number of separate items which cannot really be joined because they do not appear in a definite order. In the example below it would not be sensible to join the numbers for the different continents. The two axes could be reversed in this case.

In climatic graphs it is conventional to use a line graph for temperature and a bar graph for rainfall.

Percentage of population having Internet access, 2007

Histograms

Histograms are more complex. They look similar to bar graphs but they are used for a different purpose. They show what are technically called *distributions* or *frequencies*. On the x-axis are a series ranges of numerical values; the values do not overlap, e.g. 0 – 0.99, 1.00 – 1.99, 2.00 – 2.99 etc. On the y-axis there are the number or percentage of items of each value. The following example shows the sizes of a sample of pebbles on a beach. The modal size range (the one with the most pebbles) is 8.0 – 11.9cm.

Divided bar graphs

Divided bar graphs are used to show how a set of data can be broken down into fractions or percentages. A good example would be the percentages of different ethnic groups within a population. They have the advantage of being very simple to plot and use.

Population by continent, 2006 millions of people

Europe 728 | Asia 3958 | Africa 916 | North America 518 | Oceania 33 | South America 376

Pie graphs	Although they look very different, pie graphs can be used in exactly the same way as divided bar graphs. They require care in plotting by hand. Where the values are in percentages, to convert this to degrees, each percentage is multiplied by 3.6, to give 360°, as in the following example.

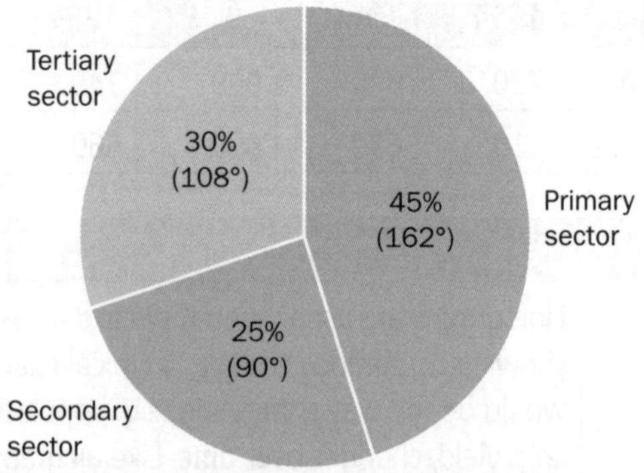

Flow diagrams

Flow diagrams are used to provide an illustration of things like traffic or pedestrian flows, population migrations or, as in the example below, world trade. They are usually based on a map but the map might be in diagrammatic form. The flow arrows might be diagrammatic as in this example, however the thickness or width of the arrows usually indicates the size of the flow and a scale is sometimes indicated for this.

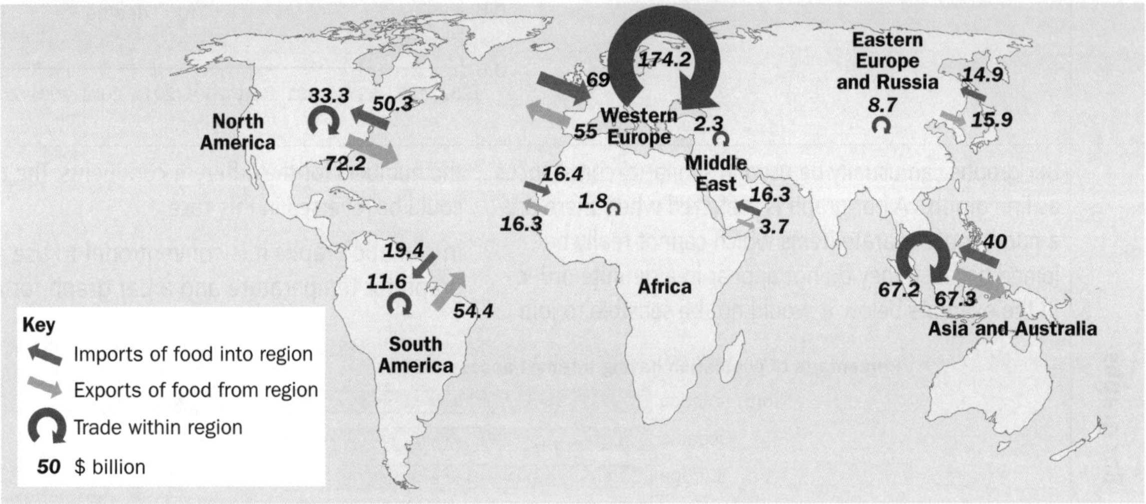

Radial graphs and wind rose graphs

Wind rose diagrams are a type of radial graph and provide a pictorial representation of wind direction. There are different ways of plotting wind roses, but all involve adding one measurement to the diagram for each day's wind direction. Some wind rose diagrams show the number of calm days in the centre.

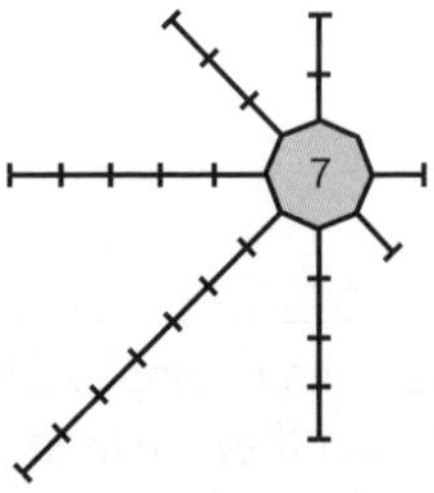

Scatter graphs

Scatter graphs are used for what is known as paired data. This is when there are two variables, in other words you know two things about a set of places. The scatter graph shows how they are related. The data is not continuous like the data shown by a line graph. There are three possible situations for scatter graphs shown in the diagram. Often a *best fit line* is drawn between the points. This does not join the points but shows the general relationship between the two variables.

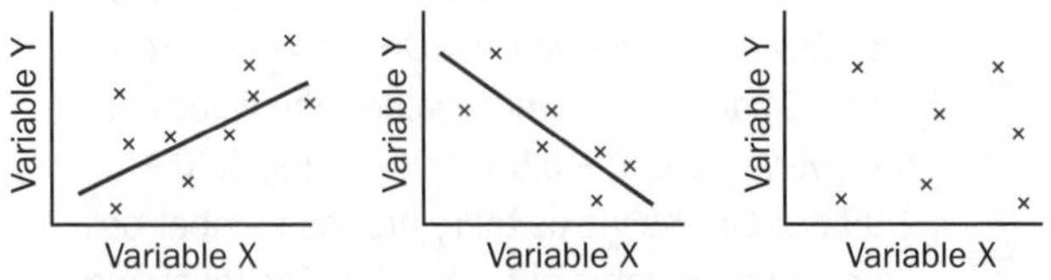

Positive correlation. When one variable increases so does the other.

Negative correlation. When one variable increases the other decreases.

No correlation.

Triangular graphs

Triangular graphs are used when you have a set of data for three variables which add up to 100%. Pie charts and divided bar graphs can show this information for one place but a triangular graph can show it for many places at once. A typical use is to show the employment structures of a group of countries as shown on the diagram. This graph reads in a clockwise direction, however other examples read anticlockwise.

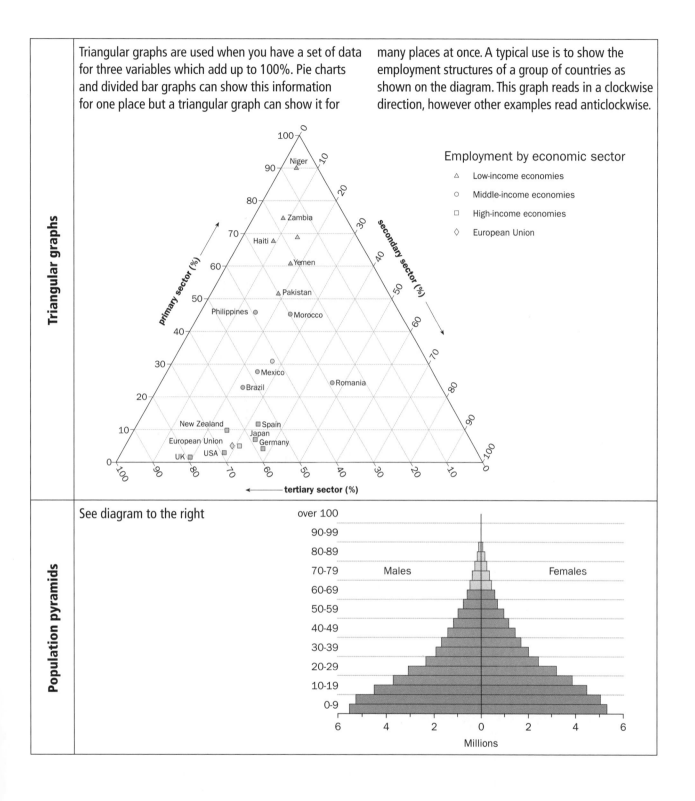

Population pyramids

See diagram to the right

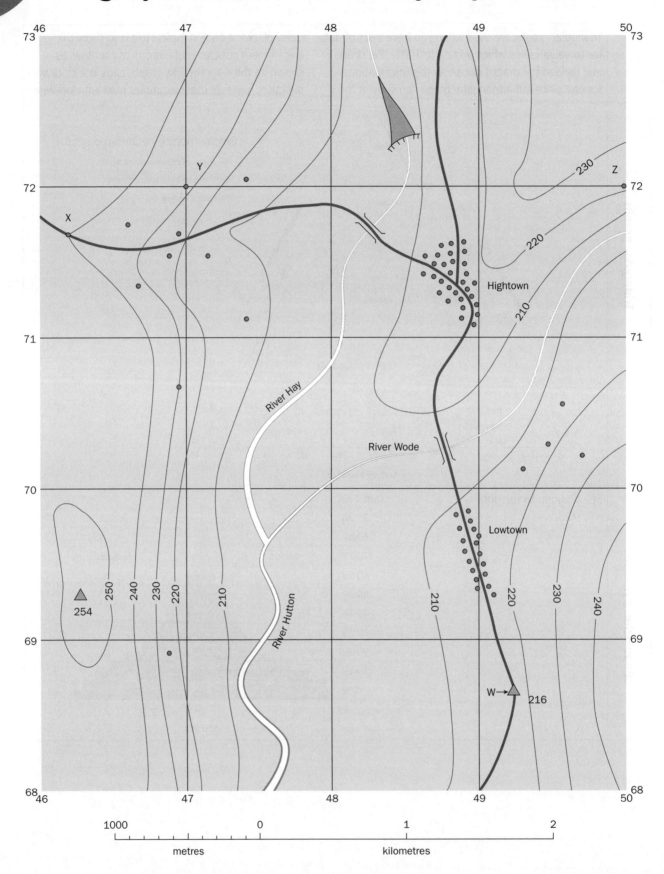

Study the survey map and answer the questions which follow.

a) i) Give the four figure grid reference of the village of Hightown. [1]

ii) Give the six figure grid reference of the trigonometrical station
on the road at point W. [1]

b) Measure the distance between points W and X:

i) in a straight line [1]

ii) along the road. Give your answers in metres. [1]

c) i) What is the difference in height above sea level between
points W and X? [1]

ii) Using your answers to b)(ii) and c)(i), calculate the gradient
along the road between points W and X. [1]

d) i) What is the compass direction from point X to point W? [1]

ii) What is the 360° bearing from point X to point W? [1]

e)

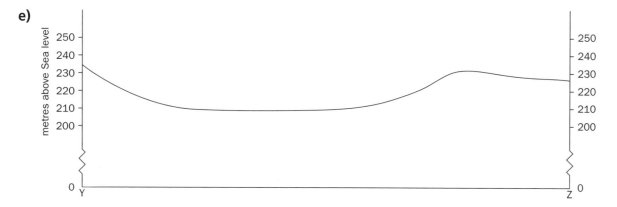

The cross section has been drawn along northing 720 from point Y
at 470720 and point Z at 500720. On the cross-section, use labeled
to arrows to show the position of:

i) the River Hay;

ii) the road;

iii) the river flood plain;

iv) the slope on the west side of the valley. [4]

f) Describe the relief of the area south of northing 700. [5]

g) Describe the drainage in the area of the map. [4]

h) Houses are shown on the map by circles.
Describe the settlement pattern on the map. [5]

i) Describe how the roads in the area of the map are
affected by the relief. [4]

[Total: 30 marks]

The topic to be investigated should be

- able to be completed in the time available by the number of people in the group.
- able to be done in an accessible and safe location.
- able to be done using equipment available.
- likely to succeed.
- a hypothesis or question.

A hypothesis is a statement that can be proved or disproved by being tested.

When researching a relationship, all other variables, such as the time of the survey, must be kept the same so that they cannot influence the results.

Health and safety considerations

The dangers of working near and in rivers (especially on outer bends of meanders), as well as near the sea and cliffs should be considered. Safety precautions which may apply include:

- Wear strong footwear and suitable clothing. Including a life vest if appropriate.
- Consider the need to wear insect repellent and sunblock.
- Do not work alone.
- Questionnaires should be conducted in pairs in safe locations.
- Contact phone numbers for home and school and a mobile phone should be taken out if possible.
- Tide tables should be consulted before working on a beach.

A checklist should be made of what is needed in the field.

Sampling

- A sample is a group selected from a larger 'population', where population means the whole of whatever is being sampled.
- It aims to investigate the smallest number which would be large enough to be truly *representative of the whole population*.
- Samples can be divided into sub-groups with different characteristics if differences are significant to the investigation, for example age groups or types of vehicle.
- To be a fair test, the sample must be chosen without bias; this means that *every individual in the population must have an equal chance of being included in the investigation.* The investigator must not choose which people should be asked to answer questions.
- The larger the sample, the more reliable the results are likely to be. A sample of 30 is usually sufficient when a relationship is being investigated.

It is important that the most appropriate sampling method is used.

- Random sampling removes bias from an investigation. Random number tables are used to decide what or where to sample.

39 26 02 11 98 55

58 07 46 60 77 04

17 83 29 32 41 36

48 65 08 93 55 69

The numbers can be read in any direction, so long as it is consistent.

- Systematic sampling uses a regular pattern or order.
- Stratified sampling makes the sampling as representative of the population as possible by ensuring that different groups or types are represented in the same proportions as they exist in the total population. It is usual to balance genders and age groups. Stratified sampling has the important advantage that all parts of an area or sections of a population are included. Significant differences can be found between sub-groups.

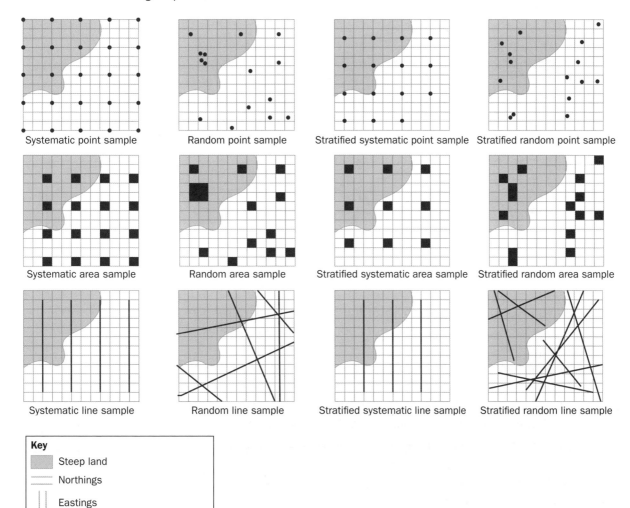

| Systematic point sample | Random point sample | Stratified systematic point sample | Stratified random point sample |

| Systematic area sample | Random area sample | Stratified systematic area sample | Stratified random area sample |

| Systematic line sample | Random line sample | Stratified systematic line sample | Stratified random line sample |

Key
Steep land
Northings
Eastings

Practice questions

1. **a)** What are the reasons for, and advantages of, sampling?

    b) Use the random number table, reading vertically down, to state which pebbles you would pick up when investigating pebbles along a line on a beach.

    c) You are taking a sample of 30 farmers to find out about the extent of organic farming in an area in which 50% are arable farms, 30% are mixed farms and 20% are pastoral. How many of each type of farmer should you plan to include for a stratified sample?

    d) i) How would you sample if investigating changes in soil depth on the valley side and floor shown on the diagram on page 138 (Chapter 33)

    ii) Describe any difficulties which could arise while doing the sampling you have chosen and of doing other methods of sampling in this area.

    e) You are to investigate changes in vegetation on the dunes shown on the diagram on page 69 (Chapter 17). What sampling method would you choose and why?

    f) Why is the choice of the sampling interval important?

2.
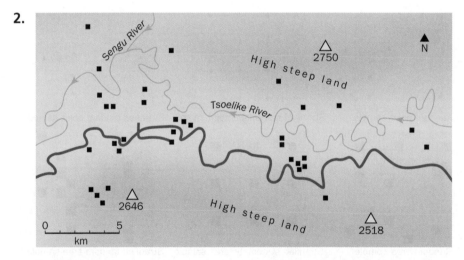

Key

------	River
▬▬	A4 road
△ 2518	Height above sea level (metres)
■	Building

Explain

i) how you would choose which 15 of the 34 householders living in this area to visit, in order to find out about reasons for the sites of houses,

ii) why other sampling types would not be appropriate to use.

3. Draw two columns, one entitled Random sampling and other Systematic sampling. Put the following descriptions into the correct column:

    - Can give a false idea by giving too few or too many counts of a feature if they are arranged in regular lines.
    - Not all features or people have an equal chance of being chosen.
    - Quicker and easier to do.
    - Ensures a coverage of the population and prevents clusters being selected.
    - Takes longer in the field because of the irregular spread of points, areas or lines.

- It is not always representative as it can miss small quantities of data.
- Not all features or people have an equal chance of being chosen.

4. **a)** What is the difference between primary and secondary data?

   b) What advantages does objective (quantitative) data have over subjective (qualitative) data?

   c) How can secondary data help in a weather study?

5. What is a pilot survey and why is it useful?

6. What should be included on all recording sheets?

7.

| Date _____ | Location _____ |

Time _____   Name of student _____

Excuse me. I am a student at X School and I am investigating the effects of aircraft noise around the airport for my IGCSE Geography coursework. May I ask you a few quick questions about this? I shall not ask your name and your replies will not be linked to you in any way.

1 Do you live in this village or town?   Yes ☐   No ☐

If you answered 'no', what is the nearest town or village to where you live?

_____

2 Approximately how often does aircraft noise annoy you when you are inside?

every day ☐   4–6 days a week ☐   1–3 days a week ☐   less than once a week ☐   never ☐

3 _____

_____

Thank you for your help and time.

Respondent's gender: M ☐   F ☐

Respondent's age (estimated):   below 20 ☐   21–45 ☐   46–65 ☐   above 65 ☐

   a) Write an appropriate question 3 with options for answers for the questionnaire.

   b) Why are boxes included for the interviewer to guess the gender and age of the person being interviewed?

8. Match the terms with their definitions.

	Term		Definition
A	Transect	1	Without order – any individual or feature could be chosen.
B	Sample	2	A statement about the subject of an investigation which can be tested and proved or disproved.
C	Population	3	A sample which divides the population into sub-groups and represents each one in the proportion that it has in the population.
D	Bias	4	A line along which samples are taken.
E	Objective (Quantitative)	5	Numerical data.
F	Systematic	6	A distortion which causes the population to be misrepresented. Often results from subjective judgment.
G	Stratified	7	A selected number which is considerably smaller than the total number that could be investigated if all were included.
H	Hypothesis	8	In statistics it means all and is not confined to people. It could be all the trees in a forest or all the goats in an area, for example.
I	Random	9	The data is collected according to a regular pattern.

Measuring accurately

Readings from digital instruments are more likely to be reliable than student readings of non-digital instruments but measuring errors can be reduced by:

• knowing how to read each instrument correctly.
• taking an average of three or more readings or measurements.

Measuring the cross profile (section) of the stream channel

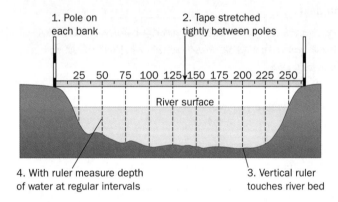

1. Pole on each bank
2. Tape stretched tightly between poles
25 50 75 100 125 150 175 200 225 250
River surface
4. With ruler measure depth of water at regular intervals
3. Vertical ruler touches river bed

The smaller the intervals between measurements, the more accurate the profile will be.

Investigating pebble load size and shape

Measuring the long axis of a pebble with callipers is more accurate than using a ruler and judging it by eye. A device called a pebbleometer can also be used.

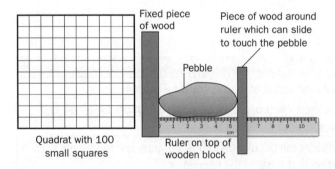

Fixed piece of wood
Piece of wood around ruler which can slide to touch the pebble
Pebble
Quadrat with 100 small squares
Ruler on top of wooden block

The average particle size at each sample point can be determined.

The angularity or roundness of pebbles is assessed subjectively by comparing them with drawings in Powers' Roundness Index Chart.

If a quadrat is used, one pebble could be collected from a set number of squares using either a random or systematic sampling method.

Counting methods

Pedestrian or traffic flows are counted either by making a mark on a tally chart or by clicking an automatic counter as each individual person or vehicle passes. A tally chart is quick to count as groups of five are made by making the fifth mark cross the first four.

TRAFFIC RECORDING SHEET

Day: Tuesday   Date: November 8th   Time: 8-8.10

Street: West Street   Site: after junction with Hope Avenue

Inbound/~~outbound~~ side   Weather: wet and windy

Mode	lorries	vans	buses	cars	motorcycles	bicycles
Tally	ЖЖ ЖЖ I	ЖЖ ЖЖ	IIII	ЖЖ ЖЖ ЖЖ ЖЖ ЖЖ ЖЖ III	ЖЖ ЖЖ ЖЖ ЖЖ ЖЖ III	ЖЖ ЖЖ III
Totals	11	10	4	33	28	13

It is important that all counts are done at the same times to be a fair test.

Each count should be long enough to give a representative sample for reliable data to be collected but not so long that students lose concentration or become tired.

Questionnaires

It is often important to do a stratified sample for questionnaire enquiries; the number in each age group asked should be representative of the total population. (Details can be obtained from Census data). For many surveys it would be acceptable to interview an equal number of males and females to obtain a balanced view.

Land use surveys

In urban areas systematic sampling can be done along one or more transects along roads starting at the town centre and ending at the edge of the CBD or town. The recording sheet may have spaces for land uses on more than one floor of a building. Land use categories are decided beforehand. Observed land uses are plotted on a large-scale plan of the area on which separate buildings are shown.

For rural land use, it is also easy to use point or area sampling, as there are usually fewer obstructions.

Vegetation

Investigating changes in vegetation species and ground coverage involves area sampling using a quadrat at regular intervals along transects. If the quadrat has been sub-divided into squares, the number of relevant squares can be counted to find the percentages.

Surveying a slope profile

Use two ranging poles, a clinometer, a prepared recording sheet, pencil and clipboard. It is important that the ranging poles are kept vertical and rest on the surface but do not sink in to the ground.

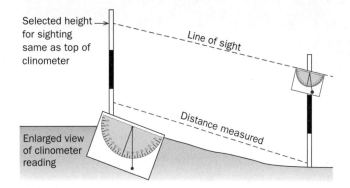

Selected height for sighting same as top of clinometer

Line of sight

Distance measured

Enlarged view of clinometer reading

Bipolar surveys

These use a range of scores to assess environmental quality, for example from 0 to 4, with 4 representing the highest quality and 0 the worst. Some scales use negative and positive figures either side of 0: the average situation is represented by 0. Negative figures show the extent of undesirable quality and positive figures indicate the extent of the good aspects of the site.

	0	1	2	3	
Very low quality of house exteriors					Excellent quality of house exteriors
Very low quality of roads and pavements					Very high quality of roads and pavements
A lot of litter					No litter

Whichever characteristics of the environment are selected to be measured, it is important that all sites are assessed at the same time as each other.

Recording observations in the field

A landscape or scene is recorded by a field sketch or by taking a photograph.

Practice questions

1. **a)** Explain how you would measure the velocity of a stream, using an orange.

   b) Look at the diagram (page 168) of a method of measuring the cross-profile of a stream channel. How would this be done differently if the stream had gently sloping banks?

2. Explain how you would investigate traffic flows on different roads in a CBD.

3.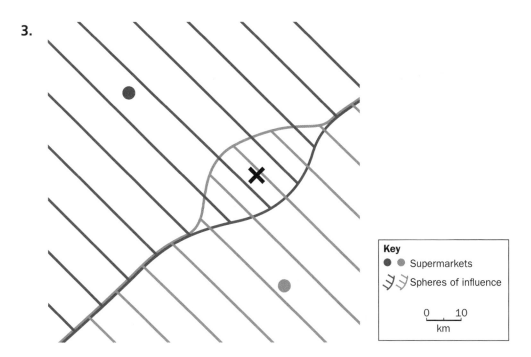

   Key
   ● ● Supermarkets
   ⤳ ⤳ Spheres of influence

   0 ___ 10
   km

   a) What does the map tell you about area X?

   b) Describe how you would find out the spheres of influence of the two supermarkets.

   c) Why would it be a good idea to have 'Do you live locally?' as the first question?

4. Explain how you would investigate change in vegetation species and height across a lines of sand dunes using a systematic method. Use the following words in your answer: count, high water mark, quadrat, ranging pole(s), recording sheet, regular, right angles, ruler, tape measure, transect.

5. a) State three characteristics that could be assessed to determine environmental quality.

  b) Different groups undertaking a bipolar survey will need to use a copy of an agreed environmental quality reference sheet. Explain its purpose.

Characteristic	Description	Score
Quality of roads and pavements	Tarred, without potholes or breaks in the surface	+2
	Tarred, with some breaks in the surface	+1
	Tarred, with many potholes or breaks in the surface	0
	Earth road with a smooth surface	–1
	Earth road with many potholes	–2

  c) What will an environmental quality recording sheet need to have apart from descriptions of the variables, a scale and spaces to tick the score for each characteristic?

  d) How could a group of students reduce the subjectivity of giving an environmental quality score?

  e) What can be done with the total environmental scores for each survey site?

6. Why should the times and day be chosen carefully when conducting investigations such as questionnaires and traffic or pedestrian counts?

7. Compare the advantages and disadvantages of using field sketches or taking photographs to record features of a landscape.

8. Explain how you would test a river for pollution and observe indicators of water quality.

Coursework: presentation, analysis and evaluation

Presentation of the data

The aims of the investigation and the hypothesis being tested will be stated. It is important to look back to this when answering questions.

Analysis and interpretation of data

Simple statistical analysis

- Rank order indicates the relative importance of a feature.
- The range of the data shows its spread or distribution.
- Mean, median or mode indicates the middle of a data set. If there is a normal distribution (without extreme values on one side which would distort the mean), the mean is a good indicator; otherwise the median should be used, as it is not affected by extreme values. The mode has limited use but the modal class of a histogram can be a valuable indicator.

Weekly total sunshine hours at a place over a nine-week period

hours (arranged in rank order)

```
       64
       60
       53          ┌─────────────────────────────────┐
       44          │ mode = 24                        │
       35 → median │ range = 21 to 64 = 43            │
       33          │ mean = 358 divided by 9 = 39.8   │
       24          └─────────────────────────────────┘
       24
       21
total  358
```

Simple statistical analysis of a set of data

Describing trends

Comment on Increases and decreases, both over the whole time period and the pattern within it. Describe in detail, such as *slight* decline, *large* increase, and support your description by quoting the amount of increase or decrease in figures or words (it 'halved' or' tripled' etc.) and the period of time over which it occurred. Always refer to both axes of a graph. Never simply list figures e.g. the values for each year in turn - figures must be interpreted and used.

Patterns, relationships and anomalies

Patterns from data in tabular and graphical form can be used to deduce relationships. Example of patterns are that no rain falls when cloud cover is 4 oktas or less and most rain falls when the cloud cover is 8 oktas, suggesting a positive relationship between the two variables.

A description of patterns on a map involves using words such as 'mainly', 'least' and 'more' while expressing the distribution of the feature in question. It may also be appropriate to refer to shapes, using terms such as 'round', 'elongated', 'linear', 'nucleated', 'dispersed' etc.

Scatter graphs are valuable for recognising whether or not a relationship exists between two variables and whether it is positive or negative (inverse).

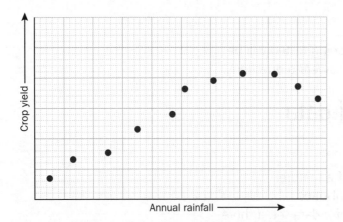

Suggesting explanations for the results

Use your knowledge to explain the distributions, trends, patterns, relationships or anomalies or suggest why none were found.

Make conclusions based on the data

Can the hypothesis be accepted?

State whether or not the evidence allows you to accept, reject or partially accept the hypothesis. Quote data to support your decision.

To what extent was it reliable?

A good sample will be as representative of the population as possible and will avoid bias. It is acceptable to describe it as a 'fair test' or 'reliable', provided that you explain why it is so, but do not describe it as 'accurate'.

Several anomalies suggest that the hypothesis is partly but not wholly true. Support this conclusion by referring to the anomalies.

Evaluating the investigation - to what extent was it successful?

If the hypothesis was sensible and the investigation well planned, it should have been generally successful. If it is only partly true, was enough data collected for the hypothesis to be tested? Was any data missing?

How could it be improved?

- Was the investigation affected by bias? The time and day chosen for the survey can give bias by preventing certain groups of people from being included.
- Were the investigating techniques the most appropriate?
- Was the sample size less than 30?
- Were the results affected by an unexpected factor?

Suggest ways in which any of the faults could be corrected. Don't be vague - the suggestion 'do more studies' will not gain credit unless the type and/or location of the extra study is stated.

One possible explanation for an anomalous result is student error in measuring or reading an instrument. The study could be improved by repeating it using better equipment, such as digital measuring equipment which would eliminate reading errors.

How could the investigation be extended?

Suggestions must be practical. Comparison with a past study or with a study obtained from a secondary source would be valid. Suggest other hypotheses about the subject which could be used to widen the research.

Practice questions

1. Look at the scatter graph showing the relationship between rainfall and crop growth (page 174).
   a) Comment on any relationships shown.
   b) If points are very close to a best-fit line, what does it indicate about the strength of the relationship?
   c) What is meant by an anomaly and how could you identify one on a scatter graph?
   d) If a close relationship is found, does it mean that one variable causes the other? Explain your answer.

2. The area has four plant species: P, Q, R and S. Four line transects were taken to investigate the vegetation in the area.

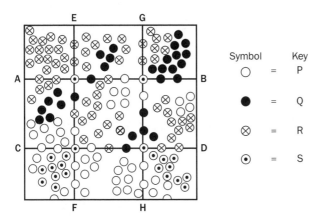

a) List the plant species on the transect line from G to H.
b) What does this transect suggest about the most abundant vegetation species in the area?
c) Use the diagram and your answer to (b) to explain why line sampling gives unreliable results in some circumstances.
d) Why should you not describe the results of any sample as 'accurate'?

3. a) What type of sampling is shown in the diagram below?

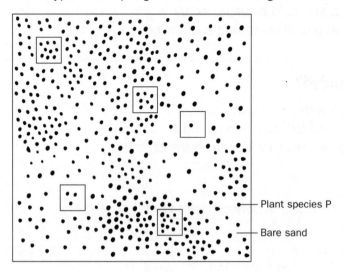

— Plant species P

— Bare sand

b) Each sample is 1 metre square and the whole area is 10 metres by 10 metres. Estimate the total number of plants of species P in the whole area. Show your working.

c) The diagram below shows the same sand dune area after it has been opened up to tourists. Estimate, using the samples shown, the new total of plant species P for the whole area. Show your working.

d) Comment on the results of these two samples compared with what you see in the two diagrams.

e) Suggest why this sampling technique failed to show the effects of the tourism and suggest a type of sampling that would have worked well.

f) List ways in which results of samples and surveys can be improved.

4. Describe the pattern of roads in London and the surrounding area.

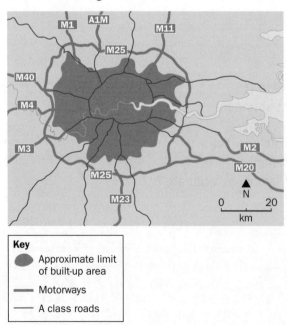

Key

⬭ Approximate limit of built-up area

▬ Motorways

— A class roads

5. Describe the trends shown by the bar graphs.

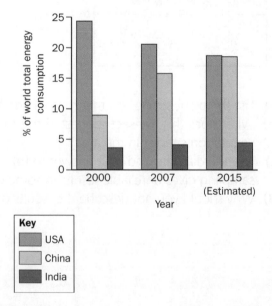

% of world total energy consumption

Year

Key

▥ USA

▤ China

▨ India

Coursework skills: Exam-style questions

1. For their IGCSE coursework, students at a school on the western side of a continent in the Northern Hemisphere decided to investigate the weather using the school's weather station instruments. They selected two hypotheses and took measurements for 20 days in February.

   Hypothesis 1: Diurnal (daily) temperature ranges are greatest when atmospheric pressure is highest.

   Hypothesis 2: Precipitation is greatest when winds are from a westerly direction.

   a) i) Name the instrument used to obtain the temperature data and explain how the students should read and reset it to get accurate measurements for this investigation. [3]

      ii) How would the students take the pressure reading from an aneroid barometer? [1]

   b) The students recorded their data in a table.

Day	Maximum temperature (°C)	Minimum temperature (°C)	Diurnal range (°C)	Average pressure (mb)	Precipitation (mm)	Wind direction
1	1	−6	7	1032	0.1	WSW
2	1	−7	8	1040	0	W
3	2	−5		1037	0.2	ESE
4	1	−7	8	1035	0	E
5	6	1	7	1034	0.3	SW
6	8	1	7	1027	0.4	W
7	8	4		1025	0	NNW
8	10	5	5	1026	0.3	W
9	11	8	3	1020	2.0	WSW
10	9	1	8	1009	2.4	SW
11	6	−2	8	1023	1.3	W
12	7	2	5	1027	0	WNW
13	11	6	5	1025	3.5	SSW
14	12	7	5	1018	7.1	SSW
15	18	11	9	1021	2.3	W
16	13	4	9	1025	3.6	SW
17						
18	11	3	8	1027	4	SW
19	12	7	5	1022	0	SW
20	12	10	2	1022	8	SW

i) Complete the table by inserting the diurnal temperature ranges
    for days 3 and 7. [2]
ii) Suggest one reason why there are no records for day 17. [1]
iii) Calculate and compare the mean daily temperatures on days
    5 and 20. [3]

c) The temperature and pressure data was presented on a
    scatter graph.

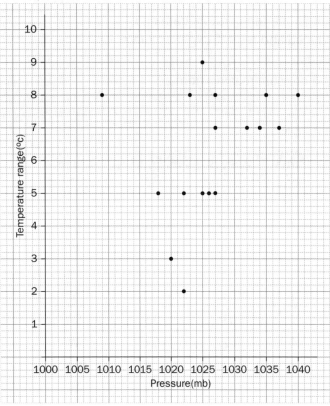

i) Plot the values for day 15 on the scatter graph. [1]
ii) What would the students conclude about hypothesis 1: Diurnal
    (daily) temperature ranges are greatest when atmospheric
    pressure is highest? Use data to support your view. [2]
iii) Explain why the students expected daily temperature ranges
    to increase as pressure increased. [3]

d) i) Explain how the students would use a wind vane to determine
    the wind direction. [2]
   ii) How should the rain gauge be sited to give accurate readings? [3]

e) To see if there was a correlation between wind direction and rainfall,
the students drew dispersion diagrams for each wind direction.

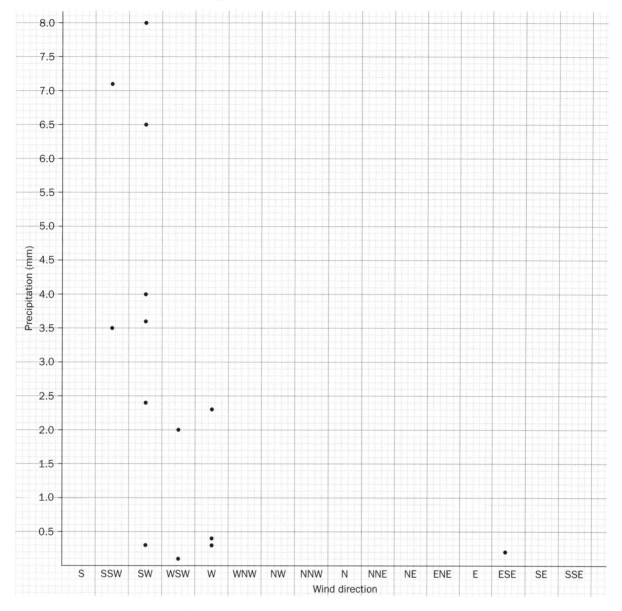

i) Complete the diagram by plotting the information for day 11. [1]
ii) Which result is an anomaly? How can it be explained? [2]
iii) Do you agree with Hypothesis 2: Precipitation is greatest
when winds are from a westerly direction? Quote evidence
for your conclusion. [2]
iv) Explain your results. (This instruction means that you should
explain why you would have expected the conclusion you
have described in (e)(i).) [2]
v) Describe two ways in which the investigation could be
extended. [2]

2. Students in a school on an island off the coast of northwest Africa studied tourism on their island by testing two hypotheses.

   Hypothesis 1: The amount of benefit to the economy from tourists varies with their ages.

   Hypothesis 2: The number of tourist visitors decreases with increasing travel time to the resort.

   The students planned a questionnaire.

   Questionnaire for tourists

   Date _____   Time _____   Location _____

   Hello, I would be very grateful if you would answer a few short questions for me to help my coursework about tourism for the IGCSE examination.

   1. Do you live on the island?   Yes   No

   2. Which age group are you in?

      Under 20   20–34   35–49   50-64   65 and over

   3. What is your home country? _____

   4. How did you travel during the longest part of your journey to get here?

      Plane   Inter-island ferry   Cruise ship

   5. How long was your flight/voyage to here?

      Under 1 hour   1–2 hours 29 minutes   1½ hours – 3hrs 59 minutes

      4 – 5 hours 29 minutes   5 ½ hours and over

   6. What type of accommodation are you staying in?

      Apartment   Caravan/ camping site   Hotel   With friends/family

   7. Have you used, or will you use, the following to travel on the island?

      Hire car   Coach tour   Boat trip   Public service bus

                                  None of these

   ---

   a) i) One student suggested that the questionnaire should be asked of people waiting in the queues to check in at the airport for their flights home. State one advantage and one disadvantage of her proposal.   [2]

   It was decided to ask 100 people, so the students divided into 20 pairs. Each pair had to interview 5 tourists, one from each age group. The teacher told them to ask the 5th person that passed after each interview had finished until they had managed to ask one person in each of the age groups.

   ii) What kind of sampling did the teacher recommend and why was it a suitable method to use for this enquiry?   [4]

   To conduct the questionnaire, they stood at equally spaced intervals on the pavements along the roads leading to the two beaches and all started their questionnaires at 10.00 in the morning. They did not ask anyone who looked younger than 18.

iii) Why did the student groups spread out and start their questionnaires at the same time? [1]

iv) Suggest why the students needed an answer to question 1 at the start of the interview. [1]

b) To test the first hypothesis the students totalled the answers to questions 6 and 7.

Answers to question 6					
Age group	**Apartment**	**Caravan/ camping**	**Hotel**	**With friends/ family**	**Total**
Under 20	9	10	0	1	20
20–34	12	7	1	0	20
35–49	12	2	6	0	20
50–64	9	0	10	1	20
65 and over	5	0	13	2	20
Total	47	19	30	4	100

i) The students presented the information about accommodation in the form of divided bar graphs for each age group. Complete the bar graph for the 35–49 age group. [2]

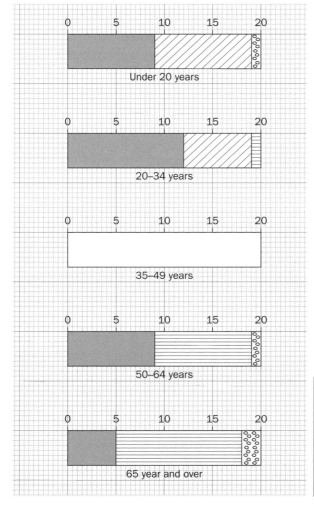

Key
Type of accomodation

- Apartment
- Caravan and camping
- Hotel
- Family and friends

ii) What general conclusions can be drawn about how
accommodation varied with age groups? [3]

iii) Rank the different types of accommodation in the order in which
they would contribute to the economy. The second highest has
been done for you.

1 _____ **2** Apartments

3 _____ **4** _____ [2]

iv) Suggest one reason why different age groups might choose
different types of accommodation. [1]

c) The students put the answers to question 7 in a data table.

Answers to Question 7					
Age group	Car hire	Boat trip	Coach tour	Local bus	None of these (stayed in resort)
Under 20	0	2	0	5	18
20–34	10	4	3	2	10
35–49	22	15	4	3	0
50–64	20	20	5	3	0
65 and over	2	14	18	5	1
Total	54	55	30		29

i) Complete the table by inserting the total number of users of the
local bus service. [1]

ii) The students decided that car hire, boat and coach trips would
contribute most to the economy, so they produced a pie chart
of their use by the age groups. Explain how they would calculate
the angle for the 50–64 age group. The total of all users of these
three transport types is 139. [3]

iii) What general conclusions can be made from the information in
the pie chart and tables about how different age groups contribute
to the economy by using transport? [2]

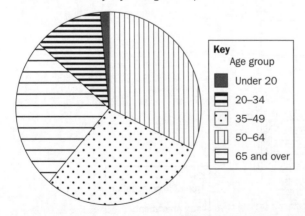

Key
Age group
■ Under 20
▤ 20–34
⬚ 35–49
▥ 50–64
▤ 65 and over

iv) Can you accept hypothesis 1: The amount of benefit to the
economy from tourists varies with their ages? Explain your answer. [3]

d) i) One student suggested using the airport's website to find out where the arriving flights came from. What type of data is this? [1]

  ii) To consider hypothesis 2, the students plotted the results for question 5 on a bar graph to see if journey time affected the number of tourists.

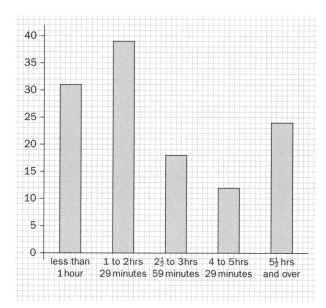

Suggest one reason for the results of the longest journey time [1]

  iii) Do you agree with hypothesis 2: the number of tourist visitors decreases with increasing travel time to the resort? Quote data to support your conclusion. [3]